SHOOTING THE NET

Since three-fourths of the earth's surface is water and one-fourth land, it's perfectly clear the good Lord intended that man spend three times as much time fishing as he does plowing.

Anon

SHOOTING THE NET

PERSONAL INSIGHTS INTO THE PAST, PRESENT AND FUTURE OF THE BRITISH FISHING INDUSTRY

MICHAEL CHARMAN
WITH GEORGE BEHREND

St. Ives from the Quay.

• MARITIME HERITAGE •
from
The NOSTALGIA Collection

This book is dedicated to the memory of my mother Nora

First published in 2006

British Library Cataloguing in Publication Data

A catalogue record for this book is available from the British Library.

ISBN 1 85794 223 X
ISBN 978 1 85794 223 1

Silver Link Publishing Ltd
The Trundle
Ringstead Road
Great Addington
Kettering
Northants NN14 4BW

Tel/Fax: 01536 330588
email: sales@nostalgiacollection.com
Website: www.nostalgiacollection.com

Printed and bound in Great Britain

Half title **Two 'old salts' at Cullen harbour, North East Scotland.** *Buckie District Fishing Heritage Museum*

Page 2 **The *Algrie* at sea.** *Don Liddicoat collection*

Title page **St Ives quay before the First World War.** *Charles Prynne collection*

ACKNOWLEDGEMENTS

I would like to thank Ray Kinghorn for his assistance in the preparation of this book, Max Channon, Frank Macleod of the Buckie District Fishing Heritage Museum for his invaluable contribution, and my niece Maryrose A. Davey for the poem.

BIBLIOGRAPHY

Anson, Peter F. *Fishermen and Fishing Ways* (Wakefield, EP Publishing, 1975; 0-71581-049-9). A reprint of a 1932 book by George Harrap.

Fisher Folk-Lore (Leighton Buzzard, Faith Press Ltd, 1965)

Gray, Malcolm *George Washington Wilson and the Scottish Fishing Industry* (Aberdeen University Press, 1982; 0-95079-092-3). Photos from the George Washington Wilson Collection.

Hustwick, Ian *Moray Firth Ships and Trade During the 19th Century* (Aberdeen, Scottish Cultural Press, 1994; 1-89821-805-6)

Pearson, Michael *Iron Roads to the Broads and Fens* (Wayzgooz, 2005)

Quill Smart, David J. *The Cornish Fishing Industry* (Penryn, Tor Mark Press, 1992; 0-85025-331-4)

Slater, James *A Sea-Faring Saga* (Fleetwood, Commercial Fishing Enterprises Ltd, 1979; 0-95020-391-2)

Stewart, Ron *Sail and Steam* (Elgin, Moray District Libraries, 1986; 0-95045-817-1)

Taylor, James *Northern Fishing Ports* (Fraserburgh, Visual Image Productions, 1993; 1-87426-970-X)

Wilson, Gloria *More Scottish Fishing Craft and Their Work* (London, Fishing News (Books), 1968)

CONTENTS

The tide is good, the day set fair,
Goes up the trawlerman's cry,
So call the lads and out we go,
Some fishing we will try.

So out they set from home port now,
The weather calm and still,
And if the luck is with them now,
The boat's hold they will fill.

The hours they fish turn into days,
And then it's time to leave,
For out on the horizon
A storm they do believe.

The sky turns black, the thunder roars,
The rain is lashing down.
Turn her to the wind now, lads,
Or surely we will drown.

For hours and hours they battle the storm
Merely to stay afloat,
For in this devilish evil storm
Their sanctuary the boat.

They love this job, these trawlermen,
It's in our blood, they say;
The danger is part of our way of life,
We would face it every day.

If they should see the dawn of day
How lucky they will be,
God will have brought them safely home
From the wild tempestuous sea.

Maryrose A. Davey

INTRODUCTION

GEORGE BEHREND

There are various types of deep-sea fishing, as there are types of fishing boat, or fishing smack as English people usually call them. The main types are drifting, where the boats are called drifters; trawling, using trawlers; seine-net fishing, introduced by the Danes to the British, also carried out by trawlers; and finally line-fishing, using short lines baited by hand with mussels and attached to the long main line; the latter can be 12 miles long in Scotland, but they are much shorter in Cornwall. It takes some time to pay them out and then wind them in. Consequently this laborious business of buying the bait, or getting it by inshore fishing perhaps, and baiting by hand is disdained by some fisherfolk such as those living in Fife, but it is practised both by those working out of Newlyn in Cornwall and from Buckie and nearby ports like Fraserburgh, Peterhead and Aberdeen in north-east Scotland, where Buckie crews often land their fish, as they are nearer the fishing grounds.

The British Government sent six civil servants to negotiate all UK fish quotas with the European Union in Brussels. The Danes sent 100 civil servants to do the same thing. Result: the Danes have a larger quota for some species of fish than the British, and many fishing boats in north-east Scotland are laid up. Some are almost new. It is said that the Lowestoft deep-sea fleet has almost totally disappeared too, and what is left of it mostly does inshore fishing. Cornish fishermen are not quite so severely restricted as the Scots, as plaice and sole are more plentiful than cod and herring.

The demand for lobsters and crabs has increased over the last 20 years or so compared to before the Second World War and after the war up to about 1960, and some Britons now eat crayfish (*langoustes* in French), which are a delicacy preferred by the French even to lobsters, and most of the catch goes to France, by sea or road from Cornwall, by road or air from Scotland. These have little to do with deep-sea fishing, more with inshore fishing with lobster and crab pots, constructed in such a way that the crustaceans are prevented from swimming out again once they have swum in to eat the bait. The demand is such that these are now farmed, as salmon is, the stocks of such fish in the wild being insufficient to meet demand. Langoustines, never seen in Britain until about the turn of the present century, are also farmed, taking advantage of global warming, which has made the water warm enough for them. They look very similar to baby crayfish, and have been eaten as a delicacy in France for years.

The fishermen of Newlyn and Buckie put up with much greater hardship, staying at sea for several days, or else rushing straight back to the fishing grounds after emptying their

Cornish fishermen – and a lady visitor – outside the Shamrock Lodge in the 1930s. *Charles Prynne collection*

hold at their home port or another of their choice, when a shoal of fish has been found. There were, however, fundamental differences between Cornwall and Scotland, due to the nature of the local fishing grounds, and the position of both in relation to London.

As a consequence of this, Cornish fishing boats, called 'smacks', tended to be smaller than Scottish fishing boats, those round Buckie being called 'skaffies', and both were pretty primitive open boats. They had a stem and stern, the planks of the sides being attached to these. But all fishing boats cost money, and throughout the period under consideration there was never enough money. The City of London lent money all over the Empire, and various other places too, like South America and Egypt, and thus built up huge resources, out of all proportion to the wealth generated in Britain itself for its own inhabitants.

Today people are so busy disparaging the British Empire that they forget how it worked before the days of international tele-communications. Rule 1 was that no natives of the Empire could come into the United Kingdom and take over trade from the British, although Indians established colonies all over the rest of the Empire except Britain; and Rule 2 was that the City of London must be powerful enough to control the Empire trade direct or through various local banks, and thus get richer. These unspoken rules were not so much as mentioned in a public school education, with the exception of Eton, perhaps. The rest were designed to turn out servants of the Empire, Army Officers, civil servants and, of course, evangelical missionaries.

Again, it was never explained in school or university, if you were not a history specialist, that the British Isles was not part, but boss of the Empire, being run by the Home Office for England and Wales, and the Scottish Office for Scotland – laws remained separate when the 1707 Act of Union was passed – while Ireland (until about 1921) had a Viceroy, like India ('roy' being Norman French for 'King', 'roi' in modern French).

George P. Uren of St Ives with his fishing boat *Stolen Residue* in about 1930, named because of a family dispute over a brother's will. *Charles Prynne collection*

So wealthy landowners in Cornwall had easy access to the banks of Bristol, connected by railway to London as early as 1838, and could pay to have their fishing smacks built, hiring people to run them. Gradually the landowners amalgamated or simply left the organising to directors, becoming shareholders, but the hired labourers received no share of the ship's profits. They were paid a wage, which varied with the catch, with a bonus for a good catch, but the ship made a profit from a good catch after paying the crew and the skipper's wages. There were plenty of hireable labourers; men were press-ganged to go to sea in Plymouth-based warships, and when they escaped that servitude after several years at sea, going fishing was a possible way of earning a living of sorts.

It was very different in Scotland. Local banks, which often went bankrupt – hence the term – lent money to the group of people recruited by the skipper, who organised the building of a boat locally; thus they were called share fishermen. All were known to the local bank and usually went fishing from one generation to another. The people who made money were the fish curers. They bought the fish in the markets, chiefly herring, engaged the daughters of fisher families, and when the herring moved in shoals, they organised the movement of these curing 'crews', as they were called, to go and stay in the nearest port, supplied the barrels and the salt, and the disposal of the filled barrels of cured herring, regardless of whether they were at home in Buckie or far away in Lowestoft.

The type of fishing was with a drift net. This is made up of a long rope with a buoy fastened to the far end, and other buoys fastened to it along its length, so it floats more or less on the surface, holding up the top of the net. There is usually more than one net, as they are heavy and difficult to manoeuvre on board ship, and they are joined by overlapping. The net simply drifts with the tide, the rope attached to the fishing boat at the near end, with, of course, a length of netless rope between it and the net nearest the boat, so the net drifts properly. The crew then wait for a shoal of fish to swim along and run into an unexpected net at more or less full speed, which drives the head and neck of the fish through the net mesh and the rest of its body gets stuck. Because the fish have been swimming fairly fast, they cannot reverse out again, so remain stuck until the crew decide to haul the nets in, hopefully full of hundreds if not thousands of live fish. It sounds easy, does it not?

First of all, there is no use having holes in the net, so everybody, men, wives, grandmothers, retired grandfathers, had to set to and mend when nets had holes torn in them, perhaps having caught on a rock. Then, in the confined space of a fishing boat, the rope had to be kept coiled in a locker, otherwise it would not pay out properly when it was cast. When cast, the weight of the rope floating on the water dragged the rest of the rope after it. The first net was attached before it went over the side, and this operation, of casting rope and net over the side, was called 'shooting the net'. The nets were kept in the fish hold until the fishing ground was reached, when they had to be dragged out of the hold on to the deck; they might be washed overboard if stored on deck. They were tied on the rope and shot over the side using a roller-board arrangement, of which details are sketchy, in a manner that allowed the rope to drift and let the net fall into its correct position. This was easier when there was no swell and the fishing boat was not heaving from side to side, of course.

The skipper decided where to shoot the net. He acted on his own judgement, having gone to sea on leaving school, aged 14 until 1945 or so, 15 thereafter, and later 16. The 'cabin-boy'

Mending nets is no longer the job of wives, grandmothers and retired grandfathers, but it still has to be done...
Ron Jenkin

– not called that as there was no cabin, just bunks in the crew's quarters up forward – had the usual job of making tea, still called 'infusing' in the 1930s, but when the nets were hauled in ('hauling the net' as opposed to 'shooting the net'), the wretched boy had to squeeze into the rope locker and coil the rope, which stank of tar; there was almost no air, and he emerged filthy, just when the crew, tired from hauling in the net and detaching the fish, were ready for a meal, as the vessel belted for port to sell the catch for a good price before too many others got there before them. The fish was auctioned in strict order of arrival.

Until their skin hardened, apprentice fishermen got boils on their wrists that were so bad they sometimes had to knock off, thus losing pay. The skipper could make up the numbers with hired crew, as in Cornwall, as there were enough unemployed, especially during the slump, hoping for a berth. Sometimes they stayed part of the crew for months, but they could not be sharers, so got less pay. The sharers needed to keep their share payments because eventually the bank had to he paid off, after which the skipper and the sharers made more money than just being on wages. They also took pride in their boat, as it was then theirs, unlike those in Cornwall, which belonged to the boat-owners.

The drift nets varied a little, according to the fish species being sought. Pilchard mesh is about 36 squares per metre, or roughly 12 per foot, of net. Herring is a bit larger, 30 per metre maximum, and mackerel larger still, 25 per metre. Line fishermen would also go for mackerel.

Shooting the net usually took place at dusk, the nets were hauled in in the early morning, and the fish were sold at auction on arrival back at port. After a brief respite they sailed again in the late afternoon, only stopping ashore and at home all day on Sunday.

Fish like herring and mackerel, which swim near the surface, are called 'pelagic'. It was sometimes possible to see a few before shooting the net. Otherwise the skipper worked on a hunch and made his decision when to shoot the net. This hit-and-miss arrangement, dependent on the skipper's judgement, might result in bumper catches or hopelessly poor ones, and much in between, ensuring that fish stocks were not depleted.

However, in mainland Europe, as usual ahead of Britain (though it would never do to say so, in Empire days), the fishermen began trawling a good while before the British took to it, near the end of the 19th century. The drifters' catches disappeared in the Moray Firth as foreign trawlers arrived, so some took to seine-netting (described below). The Firth is 40 or more miles wide for about half its length of it, so much was outside the 3-mile territorial waters limit.

In typical British manner, between the wars, when British trawlers began to deplete fish stocks for the drifters and seine-netters, they were banned in certain places, notably the Moray Firth. But foreign trawlers could not be banned, as this was a national, not international, agreement. So the Dutch and Norwegians arrived and scooped up the fish while the Scottish trawlermen could do nothing about it, and became hard up, pursued through the Elgin Sheriff Court if suspected of trawling in the Firth. (Magistrate's Courts are called Sheriff's Courts north of the border, presided over by a Sheriff, not a lay JP.) Being late on the scene, these Scottish trawlermen had turned their steam drifters into trawlers, rather than buy new boats.

The reason why trawling was banned was to protect the local fishermen's haddock lines, lobster pots and ground trammel nets. The foreign trawlers were protected by the government gunboat *Jackal*, and many local fisherman, deprived of a livelihood, had to emigrate to New Zealand, Australia and the Americas – almost another form of the notorious 'clearances'. One instance of how the *Jackal* worked occurred on 23 February 1899 when the crew of the Portgordon boat *Good Cheer* had their nets and lines trawled through, even inside the 3-mile limit, causing

£20 worth of damage. They complained to the Secretary of the Moray Firth Fisheries Association, but received no compensation.

Territorial waters, the 3-mile limit later extended to 15 miles from the shore of a country, where possible without intruding into some other country's territorial waters, were established some 125 years ago by the Conference of North Sea Powers of 1881. The conference also laid down the arrangements for registering fishing boats with certain letters for certain ports. This is a much more arbitrary affair than registered numbers for motor cars, and now the whole lot are apparently subject to the Data Protection Act. Suffice to say that all shipping has to paint the name of its port of registration, in full, on the stern, for example 'Queen Mary II, Southampton'. But fishing boats, unless second class, must be painted, in white on a dark background or black on a vessel painted white, with the registration letters of its port of registration, followed by its numbers, in a regulation size, on the sides near the forepeak or stem. The numbers apply only to that particular vessel. Second-class vessels have the number before the letters, or did so before the Second World War. Buckie ships were registered in Banff ('BF') before 1907, after that date acquiring their own registration, 'BCK', because there were so many of them. (It could not be 'BE', as that was already utilised, by Barnstaple in Devon.) Smaller ports could not have a registration of their own, so Newlyn ships are all registered 'PZ', although there were probably more Newlyn ships than Penzance ones in about 1900. (When people suggested listing all the letters, so people could go 'boat-spotting', the government department concerned had one of those bouts of apoplexy that bureaucrats have about harmless suggestions from the public. The would-be spotters should have realised that just as stolen motor cars disappear for ever with false number plates, so any smuggler might have a false set of numbers and letters about him ready to superimpose them in case he was stopped by the Royal Navy or Her Majesty's Customs!)

While drift-net fishing collects fish that swim near the surface, like herring and mackerel, trawling for fish along the bottom gets everything, but principally those that swim along near the sea bed, such as plaice, cod and haddock, which are called 'demersal'. The trawl net is about 120 feet or nearly 40 metres long, attached to two boards called otter-boards, which weigh half a ton each and are in turn hauled along by the trawler with steel ropes called warps. The warps are passed over one side of the trawler, fore and aft, so they are about 70 feet (20 metres) apart, so the net scoops up what lies between them. The trawl net is more or less V-shaped and 11-12 feet, roughly 3.5 metres, from top to bottom. The end of the net – called the cod end – is in fact open, but is kept tied up while trawling. The net is fastened to the boards by a head rope, and also by a foot rope, which is much longer. Because the net bumps over the seabed as it is dragged along, it is fitted with wooden rollers called 'bobbins', while the otter-boards have iron shoes at the bottom for bumping along the bottom. The fish lying flat on the bottom, like plaice or sole, are roused by a steel chain that is drawn along attached to the otter-boards; this 'tickles' the fish, causing them to rise just in time to be swept into the following trawl net.

After about 5 hours of trawling the net is hauled in and emptied. The warps from the net are attached to a steam winch, or electric winch in modern trawlers, which is in the centre of the trawler just below the wheelhouse. They pass over the side through 'gallows', which are horseshoe-like raised frames, one fore, one aft, with another pair for fishing from the other side of the trawler when desired. Various pulleys on deck enable the warps to reach these positions.

On the deck are raised boards, about 2 feet (half a metre) high, and after being hauled over the side, the full net is manoeuvred with block and tackle above the pounds formed by the raised boards, the cod end is untied, usually by the mate, who has to be nimble as a flood of all sorts of things, including a large quantity and variety of fish, falls out on to the deck. The squids, starfish and other things left on deck after the fish have been sorted and gutted are attractive to seagulls, so they tend to be hurled overboard quickly; the gulls are prevented from getting at the fish with this debris available until the washed fish is in the hold. The hold is about 3 metres long and has several entrance hatches. Trawlers are about 50 metres long, 8 metres wide and draw about 4 metres depth. They can stay at the fishing grounds for a week or more – unless, as is the case today, they are not allowed to fish at all.

Once emptied, the cod end of the net is tied up and the net shot overboard again, together with the otter-boards, which have come to rest near their respective gallows. Hence the expression 'getting shot' of something, meaning 'getting rid of it'.

After shooting the empty net again, the crew set about gutting the still-alive fish, no easy business on a heaving deck. After the fish have been sorted, those too small are hurled overboard. The different types of fish are put into different baskets, and after gutting they are washed, then passed through the hatch to the fish hold, which is also divided into squares or pounds with shelves. Ice, taken aboard before leaving port and stored in the forward part of the hold until wanted, is packed on alternate shelves.

Seine-net fishing is mostly inshore. Some of it is managed from beaches with just rowing boats. The idea is to surround a shoal of fish, such as pilchards, with a net, then pull both ends in, and the shoal comes too. The Danes used it before the British, but it originally dates back as far as drift netting. One end of the main warp or rope is dropped, attached to a buoy which in turn is attached to an anchor allowed

Seine-fishing for pilchards in about 1905: the fishermen are 'tucking' the fish from within the seine, and the fish are being 'dipped out' of the seine in baskets and loaded into the dipper boats. The outer seine net corks can be detected outside the tuck net. *St Ives Museum*

to fall to the sea bed. Seine nets are in the form of a bag, the ends kept open by the tide, which has to be watched. The net is shot after the vessel has sailed from the anchor in a triangular course and has reached a position where it is broadside on to the direction in which the tide is flowing. The net is attached to the rope at this point and shot. Then the vessel returns to the buoy, and both ends of the rope are hauled in together by a double rope-coiling machine with chain belts and cog wheel, the belts attached to a mechanical winch (steam-operated before there were diesels). The tide keeps the mouth of the net open, and hauling must be equal all the time so that the net is at the furthest point of the rope from the vessel.

Sometimes this is done with two vessels and two nets, joined together with a small stop net covering where they join, and shot together. Seine-netting is used for pilchards, and at one time watchers (called 'hoer's Cornish) were positioned on cliff tops where they could spot shoals and direct the fishing boats with white flags or branches of trees, helping the fishermen get the better of a large shoal. Around the 1930s this was done with smallish rowing boats and the pilchards were landed on a beach, thousands of them at one haul. In the 1930s seine-netting took over from the remaining drifters to a great extent inside the Moray Firth, though they were disrupted by the trawlers, especially the foreign ones.

PART 1
THE CORNISH FISHING INDUSTRY

MICHAEL CHARMAN

The history of fishing in Cornwall is well told by author Dr David J. Quill Smart in his book *The Cornish Fishing Industry: A Brief History* (Tor Mark Press, Penryn, 1992, ISBN 0-85025-331-4), and the following summary is included by kind permission of the author. There are, he tells us, records going back as far as 1602 regarding commercial fishing in Cornwall. They tell us that pickled and other dried fish were sent to the continent. Indeed, there are records stating that pilchards were being exported to Nantes as early as 1570.

About this time there were as many as 2,000 fishermen in the county. Because of the increase in business the market house in Penzance was opened in 1615, and boats were coming down from as far away as London to collect pilchards; there could be up to 3,000 of the fish in one hogshead At that time the whole of the port was involved in the industry, and although the living was hard, it did at least feed hungry mouths.

Something very strange was taking place in the county about this time. Pirates from North Africa raided Cornish ports collecting fishermen and taking them back to Morocco and Tunisia to be sold into slavery. One port in particular suffered the loss of 100 men in one night.

In 1859 the whole of the industry changed when the railways of Cornwall were joined to the national system. Penzance station became a veritable hive of activity at that time. Horse-drawn carts brought fish to the terminus, and more than 1,000 tons of fish were carried by rail. It was to become very big business indeed.

Between 1860 and 1870 fleets of East Coast boats began to arrive on the Cornish fishing grounds, and by 1878 there were about 100 fishing in South West waters, as well as steam drifters. A steam-powered fishing boat had been built at Hayle in 1876, and proved her worth many times when sailing vessels were becalmed, but it would be more than 30 years before Newlyn had its first steam drifter, the *Speedwell*, by which time there were 200 steam drifters on the Cornish coast. The Cornish fisherman's reluctance to change from sail to steam was not purely conservatism. The East Coast ports were near enough to coalfields to have the benefit of cheap coal, whereas in Cornwall coal had to be brought from South Wales and was expensive. The boiler and fuel supply in a steamer also took up a great deal of space, which could have been devoted to the catch.

For some 250 years Cornish fishermen observed the religious command not to work on Sundays: the spread of Methodism had strengthened this attitude, and everyone observed the strict preachings of the local minister. Only rarely did someone who had suffered a bad week of fishing give way to the

Cornish fishing in the days of sail: views of St Ives harbour around the time of the First World War. *Charles Prynne collection*

On the left a cart is waiting to haul the fish to the railway station at St Ives in about 1900. Women are counting the herring from the 'gurries' (handled boxes) into barrels. *St Ives Museum*

temptation if a large shoal of pilchards was sighted on a Sunday.

The East Coast men did not accept this local custom, and brought fish ashore on Monday mornings, so it made a good price because there was no competition from the Cornishmen. This was a cause of dispute for many years, and culminated in riots in Newlyn and Penzance in 1898, which started on 18 May and continued for several days. When the East Coast men tried to land their catches there were pitched battles with the Newlyn men; the local police could not quell the riot and 300 men of the Royal Berkshire Regiment were urgently called in from Plymouth. Although people were injured and various Newlyn men appeared in court, no heavy sentences were imposed, but the troops and police maintained their presence to keep law and order because there were still about 100 East Coast boats at sea. Eventually calm was restored and in time the problem was forgotten.

The East Coast men had always argued that to abstain from fishing on one day of the week meant a loss of earnings to the crew; their boats were expensive, and needed to be worked intensively to be profitable. By 1903 even the Newlyn boats were working on Sundays – they had reconsidered the issue.

Dr Quill Smart says that it always takes his breath away to see how things have changed in Newlyn since he was a boy. Almost all the crews are self-employed share fishermen, and he doubts if their sons will follow in their fathers' footsteps. It is certainly a fact that Newlyn faces a ruinous future for its trawlers with the present restrictions placed upon it. But one must not fall into the trap that Newlyn only has trawlers – in fact, there are many other forms of fishing going on within the port, such as gill-netters, long-liners, crabbers, and even hand-line fishing. So you can see that there is a great diversity of activity within the village: at present the market handles £20 million annually, and it is probably the country's premier fishing port.

Above On a summer's day in the 1920s boats are landing their catches from the punts to a cart at the base of the old lifeboat slip at St Ives. From here they will be weighed on the weighbridge near the 'Plat' before being sold. *St Ives Museum*

Below Punts landing mackerel at St Ives on a spring day in about 1900. The mackerel luggers are outside the harbour awaiting water in; those luggers that have 'saved water in' can be seen in the background. *Charles Prynne collection*

Above Men leaning on the railings of the fishermen's lodges look on as the fish are landed from punts by the old lifeboat slip. *St Ives Museum*

Below Fish being sold on the foreshore at St Ives in about 1900: there are conger and ling, with rays or skates in the 'gurry'. On the extreme left a fish salesman is handed a telegram by the telegram boy. *St Ives Museum*

Above Young and old look on while the fish sale takes place. The schooner in the background is one belonging to Short & Dunn, who were local coal merchants. The lady on the extreme right has purchased a ling. *St Ives Museum*

Below A long-liner's catch being sold on the 'Plat' at the top of the old lifeboat slipway in the late 1920s. *St Ives Museum*

Like Dr Quill Smart, as a small child I spent a great deal of time in Newlyn. It was a magical place for a boy to grow up. Much time was spent at the very end of the north pier fishing. Both north and south quays had been built in the late 19th century, at a time when artists such as Stanhope Forbes, Norman Garstin and Frank Bramley painted in what was to became known as the Newlyn School; more than a century later there are still artists living and making a living in the village. A new market building was opened by Diana Princess of Wales in 1988, and in 1990 a new development, the Mary Williams Pier, was opened by the Queen.

I spent many a fine day in Newlyn, a good many of then angling. Many of those times I was with a relative or two. I remember one day when I was with my brother Tony and his son Dennis, my nephew. We had walked up the north pier and set up past the third lamppost (we have all caught some fine fish there over the years). That day we were fishing with ragworm and lugworm, both laid on the ground in some old newspaper. Dennis had also brought along some plastic lugworms, bought at a shop on the way over to Newlyn from Penzance. He had taken two of them out and put them on top of the wall. We set up and started fishing. After some time some visitors came up and asked how we were doing and whether we had caught any fish. We had landed a wrass and a garfish. The lady in the party looked at the worms and asked what they were. Dennis told her and the lady said that they were revolting. After a few seconds Dennis grabbed a plastic lugworm, put it between his teeth, broke it in half and started to chew it.

'Would you like some, my 'ansom?'

The woman screamed and the rest of her party stood back in disgust. The lug had not

Cornwall has always been a popular venue for artists. This view was taken at low tide at St Ives in the early 1930s. *Charles Prynne collection*

The Mary Williams pier at Newlyn, opened by HM The Queen in 1980. *Max Channon*

managed to fool any fish, but it did the visitors.

Bill, a friend of mine who has since sadly passed on, was a barman in one of Newlyn's pubs for many years. On occasions he would take a drink or two. Sometimes the result of this drinking would put him into a mood of depression. On one occasion this happened and he said to Terry, a friend who was also a customer, 'After we close I'm going up north pier and do away with myself. I will jump into the sea.'

After the pub closed that night, Bill did indeed walk up the pier. He didn't know it, but he was being followed by Terry, who managed to keep in the shadows. When Bill got to the end of the pier he walked to the edge and looked over. Even at this late hour there were two visitors there. Again Bill looked over. This was now too much for Terry. He rushed over and threw himself at Bill, wrapping his arms around him. Seeing this, one of the visitors thought that Bill was being mugged, and set about hitting Terry on the head many times with his umbrella, so much so that Terry had to go to hospital and have six stitches.

Bill was furious with the man. 'I'll have you know that my friend was trying to save my life – now look what you've done to him!'

On one occasion I was over at Bell Rock fishing for wrass, and stood on some rocks. I missed my footing and fell into the sea, much to the merriment of all that were there. When I say fell in, I went right under the water completely, and had to drive home to my brother's house, who lived at Madron, and get changed.

I'm sure that my late sister-in-law thought that my brother and I were completely mad when we would go off for the whole night, fishing for sea bass at Porthcurno. On the way home we would often be stopped by the police, who would spot the rods in the back of the car.

'Had a good night, lads?' they would ask. For all they knew we could have been burglars.

I had made contact with the Royal National Mission to Deep Sea Fishermen some time prior to my visit to Newlyn in January 2003. At that time the Mission was being run by Keith Dickson, who was Acting Superintendent; he had not long been in Cornwall, having been in Grimsby in his previous posting. His assistant was Ken Bassett, a St Ives man. In truth I saw much more of Ken than I did of Keith; the latter seemed to be always heavily employed.

I was to meet only one unpleasant person during the preparation for this book, and that was at the Mission: this man told me in no uncertain terms that he thought I was 'on the make.' He also scoffed when I told him that I had promised to donate 10 per cent of the gross amount of my fee for writing the book to the Mission. All the other people that I met in Newlyn were charming and could not have been more helpful. For the single exception I have one thing to say: 'An open mind and a closed mouth make a happy combination, my 'ansom'.

I was given the Nora Bolitho Room to use while I was doing research for this book. It was comfortable with a table and many chairs, and was more than adequate. Ken made me very welcome, and I felt even more at home when I was given a large mug of coffee by Keith! Soon I was introduced to a fine man by the name of Tony Woodhams, the Chairman of the Mission, who recommended men to whom I should talk about the past and present of fishing out of Newlyn. One was Don Liddicoat.

DON LIDDICOAT

'I couldn't come ashore'

Don was quite a man, standing 6 foot 1½ inches tall and some 18 stones in weight. Known to all as either the 'Gentle Giant' or 'Big D', you only had to look at his hands to see that he was used to very hard work. A look at his arms confirmed that some of his hard-

Don Liddicoat. *Michael Charman*

earned money had been put to good use: the tattoos were works of art. His longish hair, with just a few flecks of grey, was well suited to his image. He smiled a lot, an easy-going man, an honest man, and quite happy to take whatever life threw at him. But a man very committed to his way of life. He had a great sense of fun: some of the pranks he played on others were hilarious, and those played on him were remembered with gracious humour. He told me that he was between 14 and 15 years old when he first went to sea.

'It was on the gill-netters with John Thomas on the *Boy Gary*. During the summer holidays, every time I came in I would help land the boats and get pocket money that way.'

Had he had a happy life at sea?

'Yes, I've enjoyed ever minute of it.'

Might he have chosen another profession?

'No. I adore the sea, I respect it. How can I put it? After being on shore for two or three days, I just want to get back out to sea all the time.'

He was forthright on the subject of quotas:

'Devastating, absolutely devastating. It's got to the stage now where they give you your quota and you basically go out and catch in a day or a couple of days if the fishing is good.

The *Algrie* from Stevenson's fleet, 26.20 metres long, out at sea with nets in place for a good day's fishing. Built in 1968, she is powered by a 730hp Stork engine. *Don Liddicoat collection*

Once you've caught your own quota you've got to go out and buy quotas, which is going to cost you money, even before you go to sea, and it comes out of the crew's wages. At the end of the day you're paying to go to sea to catch fish.'

Now being a skipper himself, what did he think of the monetary reward for the responsibility he carried? Did he think it was a fair reflection?

'The responsibility is too much. You're looking over your shoulder. The paperwork involved with MAFF [the Ministry of Agriculture, Fisheries and Food, more recently part of the Department for Environment, Food and Rural Affairs], if it's not up to date when you return to harbour or even at sea when the Royal Navy boards you it can actually…'

He sat forward with his head in his hands and gave a loud sigh.

'How can I put it? My God, it does make life tough, but also you're looking over your shoulder. You are not keeping a proper job, because you're afraid of what you've got to do. It's the crew on the deck that will suffer because your mind is not with you all the time. You're thinking, is that done right or is this done right? I'll have to mark the fish off the tally sheet. If it doesn't match up with the tally sheet placed with MAFF when you land, they come down on you and fine you straight away. At the end of the day, for the responsibility you get and what you get out of it you're better going as a mate or a deck hand on another boat. No worries. Let someone else have the worries.'

After a long silence Don lit another 'roll-up' thoughtfully.

'There were many very good days, all now long gone.'

I asked him about the present arrangement with his boat – was it hired from Stevenson's?

'We're share fishermen. We don't hire the boat from the Stevenson family – we actually work for them though we are self-employed. At the end of the day, when you go to sea and start catching fish you are on wages. The split of the gross is 66 per cent/34 per cent – 66 per cent to the owner, 34 per cent to the crew, with no expenses – the owner buys all the boat's insurance and fuel. When I have four hands and I work as engineer as well I get another 1 per cent – I'm on 10 per cent or 10½ per cent. Normally if there are five hands and the proper engineer is there, the share would be 8 per cent skipper, 8 per cent engineer, 7 per cent mate, 6 per cent deck hands, and 6 per cent cook.'

Although I had only been in Don's company for a short time, nevertheless I felt that this was a man that a person could trust. I would feel quite happy to put my safety in his hands. The true measure of a man is the height of his ideals, the breadth of his sympathy, the depth of his convictions, and the length of his patience. Don came across as having all of these, in good measure. I went on to ask him how long would a trip be.

'It depends on the weather. Like today, you've got a south-easterly gale of wind. So if you were in the Bristol Channel you'd get right under the 12-mile limit. You could work your way up to force 7/8 no problem. The wind comes over the land. Down here we haven't got the land, it's all open. The bay has shallow water, the sea gets whipped up and gets higher and higher there – you can't fish in the bay in a south-easterly. The Channel is no problem before Christmas in southerly and south-west 8/9. The only weather that will stop us is a 9/10 from the south-west or north-west. You don't think about working that day, you just get the gear on board.'

When on board, he presumably had to trust his workmates completely?

'You have to know what you're all doing for safety reasons. If you've got a good crew on the boat you can count on them. It's the men on the deck that work the boat. I'm in the wheelhouse. The boys keep that boat fishing. I've got to give my crew respect – if I don't give them respect, I will get no respect in return. If this is not done you land up losing a good crew.'

I could hardly believe that Don could carry all this worry and responsibility and still make light of almost anything. He told me of one of many humorous incidents.

'It concerns a guy who has never been a fisherman in his life, and he will never be a fisherman. But to his credit he is a tryer. He came to sea with us and did three or four trips. On the last trip he had a gurnard spot [a septic spot caused by the spike of a gurnard fish] on his finger. It started to swell up with poison. We had to return to Newlyn because of bad weather. I said to him that he would have to go to hospital or I could do it for him. I said that I was qualified – when you get your skipper's ticket you have to go and do your medical training. He took all this in. I said that I'd go and get the medicine box up from below. I returned with a scalpel, a magnifying glass and some bandages. He looked at this and asked what I was going to do with that lot. I said that I was going to do the job for him, that I could stitch him up, a proper good job it'd be – my ticket said that I could do it. No such thing. He put his hand on the galley table. Are you going to give me an injection, he asked. No, I said. You see that bit of wood there? Bite into that. The scalpel had gone green, so I wiped it on my sleeve. My God, he said, you're not going to use that, are you? The blades for the knife were all sealed in packets. Are you going to do it now, he asked. Yes, I said – one little nick and all the pain and pus will be gone. He put the piece of wood in his mouth and bit on it. He turned his head away to the after end of the boat as if I was going to cut him. I wasn't going to do anything to him. I got the scalpel with no blade in it and pushed as hard as I could with it. I'm sure he thought I was going to amputate it. He jumped up and hit his head on the deckhead. Myself, the mate and one of the deckies were in the galley. I was curled up with tears running down my cheeks. He never saw the funny side of it till we reached the harbour. He went up to the hospital and it burst itself – it must have been the pressure I'd put on it.'

Character is what I could see, in great quantities, grown in the soil of experience with the fertilisation of example. I would gladly have put my life in his hands – I say that with no hesitation at all. He came across as very genuine, though perhaps he sometimes let his heart rule his head! I was sure that over the years there had been more than a few who had tried to take advantage of him because of his outgoing nature.

I asked if he had any relatives at sea.

'My stepfather finished here two years ago. My brother is the skipper of a tall sea fisheries boat called the *St Piran*. He was fishing all his life since he left school. At the end of the day he had his own boat in partnership. The partnership broke up, and he went fishing by himself. I did lose a brother, Paul, at sea in 1980. He was 16. They were steaming down from the north-west coast off St Agnes on the *Geallachid Millis* [a Gaelic name translating as *Sweet Promise*]. On that April night the sea was like glass. They were in the wake of another boat, which made their boat lurch and he slipped on a piece of mackerel and banged his knee. He pulled himself up but she lurched again and he fell. He must have broken his knee. She lurched yet again, he took the weight on his leg and over the side he went.

'My eldest brother Shane saw him go over the side. They did a Williamson's turn [turning the vessel round so that it returns by the same track], came round and saw him in the water. They shouted 'Ahoy!' three times. Tried gaffing [hooking] him. No good. My brother was going to dive over the side but the skipper pulled him back – it was too far to the water. He was told by the skipper that he was not going to lose two lives, especially from the same family. That was that. We haven't found him to this day.'

I have been told that courage is being the only one who knows you're afraid.

I asked Don if he was married. I have always thought that you can see something of a person's character by observing how they smile. Don was now beaming.

'No. No. I'm not married. Single I am, but I have three children, a son of 17, another son of 12 who wants to be a fisherman – he's got a hole in his head – and a daughter of 4. She's the apple of Daddy's eye, she is. She can wrap Daddy around her little finger.'

What about religion?

'I'm a Catholic and I do believe in God, but when I'm out there I never pray. But there have been times in bad weather when I look up and I think, my God, please let us get in. I wouldn't go to church – I'm not a practising Catholic.'

I know a lot of people who go to sea that have the same point of view – fishermen are all so near to nature itself, it almost draws you to believe in God. Did he hope to stay fishing until he retired?

'Yes, I couldn't come ashore – I love the sea too much for that.'

Don Turtle, a sprightly 77 years old, has been at sea since 1941. After serving in the Merchant Navy he left to take up fishing, and he was still fishing when we spoke in Newlyn. *Michael Charman*

DON TURTLE

'I've seen millions of pounds floating in the sea'

To look at Don Turtle you would not automatically recognise him as a fisherman. With his soft voice and his gentle way, he could be taken as an accountant or a kindly schoolmaster. Even before I really spoke to him I thought that here was a man who had had many problems over the years. An old man once told me that to work out our problems we need to add love, subtract hate, multiply good, and divide between truth and error. How right he was! I started by asking Don when he had first gone to sea.

'At a very early age, at school. I was delighted when I was accepted for nautical school, to train as a seaman. At the age of 14½ war broke out and seamen were getting knocked off left, right and centre. At 15½ I finished my training and found myself aboard a merchant ship from 1941. I continued in the merchant service until about 1954, when I left and came fishing.

'The fishing when I was first engaged in it and the present day are two different things altogether. You were allowed to do what you wished to do – there were no laws and rules and there was no bureaucracy of any kind. But when the merchant fleet ceased to exist under the British register, the Ministry didn't have any revenue coming in from the merchant fleet, so they shifted it all on to the fishing. You had to pay them a fee to get a survey and a fishing certificate.

'As the years progressed and we joined the EU a tirade of bureaucracy swamped us. Almost overnight they started bringing in laws and more and more laws. One of these was that you had to phone in the time when you were landing so that Ministry officials could check your catch, etc. Well, one day I could see a scene looming up on the market with the crew. My son John, on three different occasions, had been at sea for ten days in atrocious weather. Because of the type of fishing you do, you have to work the tide and, come hell or high water, when that tide comes around you have to go, or you will lose a month. They had just come from fishing in that dreadful weather, and said that they were going to land at a certain time., but you can't always be accurate. Well, he came in and saw the officials on the quay, and since the exercise is that the officials are there to catch your fish and make sure you're not concealing anything, he thought it was all right to land the catch with the officials there. He didn't stay, but went straight back to sea to retrieve his gear as a storm was brewing up. That was considered an offence. Now there are two other offences similar to this, and it's called being out of time. He said he would be there at a certain time and he was either early or late, so he was out of the window of time, which we pleaded guilty to. In the court the prosecution was brought against us by the Ministry, but we thought it was such a trivial thing and thought nothing more about it. Then 12 months later – yes, 12 months – we got a court summons.'

Don gave another sigh and his chin fell to his chest.

'I explained to the magistrates – there were two women and a man – and the prosecution counsel for the Ministry stood up and said the offences could carry a £50,000 fine. This was ludicrous – that sort of fine was for someone trying to conceal the catch. However, he went on to tell me that they were just out of time.

'I explained what the situation was and said that I relied on their discretion to see what the offence really was. I said that we were guilty – we were out of time.

'They were deliberating the case for well over an hour. Time went by and the prosecuting council asked the clerk of the court what difficulty they were having.

'When they came out a woman addressed me and said, 'We are fining you £1,000 on each of the three counts.' To me it was like being in a Monty Python sketch, it was so ridiculous.'

I asked Don if they had given them time to pay.

'Yes. We are still paying it now.'

I am sure that there is justice for all, but it doesn't seen to be equally distributed. I then asked him what sort of boat he had.

'It's our family boat – just 74 feet.'

How far did he go out to sea to catch fish?

'Sometimes 200-300 miles.'

With a crew of five, I guessed it was a fully qualified trawler.

'No. We're netting – we've never done trawling. We shoot 8 miles of net.'

Don agreed that his life at sea had been hard, but there had been unusual moments.

'We were approached one time to do a film of "Dr Who and the Pirates". One evening when we came in from fishing, a pile of cartons had come down plus a lorry-load of timber and stuff. They built the bow up like an old galleon. Next morning all the stars arrived, some of them well known, and then this chest of swords and pistols came down to the boat. We set off to film. When the boat began to get in the seaway and roll a bit they all started to get sick. We got to a lee as best as we could near Penberth, but as they were doing their filming quite often they would break off and be sick over the side.'

A more serious event took place while Don was fishing for mackerel off Falmouth.

'The weather was not too good, but the boats were out. We were hand-lining for mackerel. We used to take a run up and drift down with the tide, and when the mackerel stopped and we came to the end of the drift we would start up again and drift down again.'

'It takes about half an hour to steam back up again from the start of the drift. Everybody will go down and have a cup of tea or whatever while we are steaming up. There was a chap on board called Jan the Pole – he was Polish, a sort of war refugee, had had a bad time with the Gestapo. I think he used to get blackouts. We noticed he was missing – he had disappeared during that steam up. Nobody knew what had happened to him. We did find signs of someone having been sick over the side. We thought that the boat must have rolled and knocked him over the side. Nobody saw anything and he was never seen again. I had the terrible task of telling his wife and kids.'

It is a sad fact that all fishermen know somebody who has been lost at sea. Fishing is the most dangerous job that you can have in this country, even before mining. In the 19th century Cornwall had them both. They were the 'good old days'. The reason that people remember the 'good old days' is that there were so few of them. Life in those days was hard.

I asked Don if he had his own boat.

'I have always owned my own boat.'

What about family?

'I have three girls and a son. All my girls are in the medical profession.'

And what of his short-term view of the fishing industry, especially in Newlyn, the quotas and such?

'Quotas do not work. Well, this is a thing that we must get sorted out, and I can't see it being sorted out with us being in the EU. The Common Fisheries Policy is a disaster. It is only those who work in the fishing industry who understand it. You won't get civil servants or scientists to understand it.

'I've had a hard life. The best part of it is being your own boss. But you've got to be able to make a living. I will just say this – I've see millions of pounds floating in the sea and never been able to get a penny from it.'

Don came over as a kind and gentle man, and there are not many of those left.

RON JENKIN

'Thirty fishermen this time next year will not be with us'

I first made contact with Ron by telephone. A friend thought that he would be the sort of person that I should meet. He was right. Even before I met Ron in person I felt that I already knew him to some degree. On the telephone he sounded outgoing and friendly, so I was very much looking forward to meeting him. I

Ron Jenkin, born and bred in Mousehole, is keenly interested in safety. *Michael Charman*

was not disappointed. He was tremendously interested in anything to do with the fishing industry. Something of a person's character may be discovered by observing how he smiles. Ron smiles a lot.

I already knew that Ron had done various jobs prior to going to sea at the age of 25. He had worked in Cheapside in London for a counting house, and after that he did his two years of National Service. When that was complete he went on to work for diary farmers under the Cow & Gate label, and from that to selling insurance. With all that under his belt he returned home to Cornwall. I knew that he had been a keen rugby player, so my first question was did he know Richard Sharpe, who played fly-half for both Cornwall and England, and had been a great hero of mine when I was a schoolboy at Lescudjack School in Penzance.

'Yes, I did know Richard, but I never played

against him. I'm a Mousehole man born and bred, and when I wanted to go to sea I had to go down and ask my uncle for a berth; as my father was part-owner of the boat, it was something of a foregone conclusion. But it had to be done. I stayed with father till he retired in 1964. Then my cousin took his father's part of the boat, and I took my father's part, and we worked like that till May 1968. At that time we had the *Spaven Nor* built down at Porthleven. It was a long-liner – we did long-lining in the summer and trawling in the winter. A little later we paired up with a boat called the *Galilaean* and went pair-trawling for mackerel and pilchards. There was already one pair working out of here and we were the second pair that went mid-water fishing or mid-water trawling down this part of the world.

'I worked for the Stevensons during the time that the *SDJ* was being refitted (this was a beam trawler – a former Dutchman – under the partnership of Stevenson, Downing and Jenkin). I enjoyed every minute of my life at sea. I became a skipper and got my qualifications. Then the doctor and I decided it was time for me not to go to sea any more. I was ashore for about a year, not doing very much except taking up golf. Then one day out of the blue I had a phone call asking me if I would like to go down to Falmouth and teach at the marine school, teaching school-leavers,

SDJ (PZ47) was a beam trawler operating under the partnership at Stevenson Downing and Jenkin. Here she is 'in full flow'. *Ron Jenkin collection*

Above *SDJ*: dinner is served. *Ron Jenkin*

Below *SDJ*: a good week's work. *Ron Jenkin*

new intakes into the industry. I had my interview with a gentleman called Dawson Fisk who was well known in training in the western region, and he gave me the job. I did that for a couple of years, but due to the lack of newcomers into the industry that faded.

'Dawson asked if I would be interested in teaching on the skippers courses, and when I asked him what he wanted me to teach, he said ship management, and all the practical stuff and meteorology. I had never done meteorology when I took my ticket – it wasn't necessary – but all he said was here's the book. Since 1990 I've been involved in training in various capacities.'

Did he intend to go on after he reached 70?

'I will go when I feel like it. I dread to think of the day that I will not be able to do anything at all.'

I presumed that he had known all the boys who had lost their lives in the Penlee disaster of 19 December 1981, when the lifeboat *Solomon Browne* was lost.

'Most of them. Actually one of them was one of my crew, Barrie Torrie. I don't think the village ever got over it. The lifeboat is out of the village now and is here in Newlyn. That was a terrible day. Strangely enough, that particular night we had gone to bed and left the key in the door. My brother-in-law, who worked on the trains – he used to work for the Post Office sorting the mail up to Bristol and back at night – heard the news and came up to our house and walked in and came into our bedroom and told us the lifeboat was lost.'

A friend of mine, who lived in Florence Place, said how he walked into that narrow street and the sea was going over the tops of the houses like a sheet – he stood under it and did not get wet.

'Horrendous, the weather was absolutely horrendous,' agreed Ron. 'I have talked to one or two people and they said they couldn't stand up because of the wind. Horrendous.'

I expressed my opinion that trawling was the most dangerous job in the country.

'That's to do with the course that I'm running at the moment – safety awareness.

The Penlee lifeboat *Solomon Brown*, lost with all hands in December 1981. *Antony Hosking collection*

Thirty fishermen this time next year will not be with us. That's one every 10 days, or one in 750 fishermen. All for a basket of fish.

'The industry, the owners, the skippers, the industry itself are very safety-minded, but it is the environment in which we work. We are no less safety-minded than other industries, more so probably because every fisherman before he goes to sea has to have a one-day course on sea survival, a one-day fire-fighting course, and a one-day first aid course. That's before he is allowed to go aboard a boat. The skipper and the owner can be fined for not ensuring that the crew have these safety qualifications.'

I could see that Ron was very serious about this subject. By looking at him you could see that anything he touched on would have his full attention. He was a man that took his responsibilities seriously, a man to be trusted, a fair-minded man. Service to others is the price we pay for the space we occupy in this world, and Ron had an abundance of service.

I said that I understood there to be very little humour on the boats today because of the situation with the quotas and the rest. It was some time before Ron replied. His words were chosen carefully.

'It is very difficult. When I went to sea first you went to sea to catch as many fish as you could in the shortest possible time. That was the whole idea. Now a fisherman has got to stick to the area that he allowed to fish in, catch the amount that he is allowed to catch in those areas, and so forth. It is such a complicated affair now. It is very, very difficult for them. There is a lot of paperwork – they have to complete fishing logs, saying how much they have caught each day, putting it down then totting it up. They are only estimated while they are at sea, but they've got to marry up as near a damn it when they get in.'

I could see how very relieved Ron was that he was no longer a skipper, and that he had that heavy responsibility removed from his shoulders. But he also understood that every survival kit should include a sense of humour.

'We were at sea one night and the weather was not very nice and I heard the sea break and I shouted to all hands to hold on and I turned my back and held on to the rail around the galley. The next thing the cook was sliding along the deck past me. I grabbed him by the scruff of the neck. As the water cleared he stood up, swore at me and said, 'Do you realise I've just lost my hat over the side?' This was just a reaction.'

I asked Ron how far they went out when he went to sea with his father.

'The furthest I have been is 120 miles past the Scillies west-by-south – we were lining out there.'

Had he ever fished in the 'Irish box'?

'Yes, that was the deepest. I have never been down the south of Ireland. You tried to get your living as close to home as you possibly could.'

There is a island not more than a quarter of a mile off Mousehole. Was I right in saying that during the Second World War the Germans used to bomb it, thinking it was a submarine?

'That's St Clements island. They did bomb it, as they thought it was an aircraft-carrier. It had what they call the 'pepper pot' on its western end, a flat base with a 'control tower' at one end. When we were young and living in the village you swam over to it. That was the way of proving yourself – I first did it when I was 12.'

I also asked Ron about the *Torrey Canyon* disaster on 18 March 1967, when that oil-tanker hit Pollard's Rock in the Seven Stones reef between the Scilly Isles and Land's End. She ripped open six tanks, and it was the first such incident to attract universal attention to the dangers of dispersants. On 28 March she was destroyed by the bombs of the Fleet Air Arm.

'Yes, I remember it well. A lot of the boats were fitted out with pumps and sprays and barrels of detergent. We were all steaming

SUMMER HOLIDAYS IN 'MOUZEL'

Do you know that when I was boy it never rained in 'Mouzel' during the school summer holidays, because if it had we would never have had the time to do all the things we did!

Take a typical day. After breakfast, run any errands, such as to the farm for milk, to the butchers or to the village shop. Next, serve out any punishment time working in the garden (the time spent gardening depended on the severity of the crime committed). I think that affects me even today, as gardening is still like penal servitude. When that was over I would run down the garden path, kick off my shoes inside the gate and make my way to the harbour.

Mouzel is built in a semi-circle on the side of the hill like a big amphitheatre, with the harbour as the arena, and that was where we all met. A lot of us had punts, propelled by sculling with one oar over the stern. Many hours were spent either fishing for mackerel and bream or cruising along the coast. When the tide was in, the harbour was our swimming pool, and when the tide was out it served as our cricket pitch, being the only piece of level ground available.

Traffic was not a problem – I can recall only three cars in Mouzel, and two of those were taxis. The buses only came through the village every 20 minutes, so all other games were played in the streets. We had go-carts, hoops, marbles and various ball games – these all came round in cycles depending on what had been decided the night before. There were, of course, the usual children's games, like hide and seek, either individually or in teams, which we called 'touch clock', when you had to stay inside certain boundaries within the village and had to remain unseen for a period of time.

I recall going back to school at the end of one such holiday and was having tea in the dining-room when in walked the headmaster, who announced that he wished to see me in his study immediately after tea! What had I done? To be in trouble only a few hours back at school was a record, even for me. I duly arrived at his study and received a lecture on hooliganism. It seems that he and some friends, all members of the Old Cornwall Society, had been invited to visit the Keigwin Arms, a private residence now, but an historical site within the village.

The Keigwin Arms, you see, was the only building left standing after the Spaniards invaded Mouzel on 23 July 1595. The villagers had been selling food to the Spanish fleet for a long time after Sir Francis Drake had defeated the Armada some seven years before, so when a law was passed preventing this trade the Spaniards attacked. They razed the village to the ground and got about a mile inland, where they burned the church of St Pol de Leon. Scorch marks can be seen on the pillars of the church to this day. Legend has it that they retreated when they mistook the ladies of the village, dressed in their red petticoats, for the army.

My headmaster and his friends had been walking through the village when they had almost been in collision with a group of hooligans, and he was ashamed to say that one of those hooligans was also one of his pupils. The one thing I remember well about this lecture was the way he described this group of boys – he said that 'they were unshod and only partially clad'.

We wouldn't have dressed like that if it was going to rain, would we?

Ron Jenkin

around the bay finding patches of oil and spraying on detergent. I remember that we had to take a lad from the Army with us to use the pumps.'

I then asked my usual question about religion. 'Do you believe in God?' He took some time before replying.

'Very much so. I am sure there is – you can call it Him or whatever you want, or whatever you like. There is some power that we don't understand. I'm not a church-goer as such, although I was brought up as a strict Methodist. I do believe there is some superior power somewhere, and whatever name you want to put to it is up to you.'

Would he ever leave Mousehole village? Property prices have risen there dramatically. On my recent visit I had seen a two-bedroomed cottage being advertised for something like £300,000. I waited with interest to see what his reply would be, as I knew that he was a born and bred Mousehole man.

'Obviously Mousehole has become a residential area – unfortunately it is no longer an industrial village in the sense that it was years ago. Now it's a popular little village, but unfortunately it is only half-occupied during the winter months. I don't want to move away from here – I would rather stay in this part of the world.'

PATRICK ('PATCH') HARVEY AND ANDREW PASCOE

'We've grown up with it'

I knew Patch's father, Edward, from the time that we worked together on the dining-cars when we worked for British Railways. But I must confess that I knew his Uncle Ben much better than his father. Ben and I worked on the same train for years; we went through the rites of passage together with several other boys who also worked on the trains. In those days I always found Ben to be happy and cheerful, and I remember back in the 1950s, when Ben had a TB scare, we all went running off to see our doctors. Roy Hosking from Newlyn, Johnny Watkins, and Jimmy and Tony Nichols were just some of the lads that we worked with.

It was a cold January day when I walked into the RNLI station at Newlyn, flurrying with snow and with a very cold east wind blowing. I made myself known to the Coxswain, Neil Brockman. I asked him if he was related to Nigel Brockman, one of the eight men who had lost their lives in the Penlee lifeboat disaster. He told me that it was his father. That was that, he said no more. The other seven men who lost their lives on that dreadful night were the Coxswain, William Trevelyan

Neil Brockman, Coxswain of the Newlyn lifeboat. *Michael Charman*

Andrew Pascoe (left) and Patrick 'Patch' Harvey have been friends since the age of 4. They work gill-nets out of Newlyn. *Michael Charman*

Richards, James Stephen Madron, John Robert Blewett, Charles Thomas Greenhaugh, Kevin Smith, Barrie Robertson Torrie, and Gary Lee Wallis – men that will never be forgotten. Bless their memories.

Neil was accompanied by two other men, Andrew Munson, who as well as being involved with the RNLI was also the Harbour Master at Newlyn, and Tim Ambrose. We were later joined by Patch Harvey, making a group of four lifeboatmen.

I asked them about the current lifeboat, *Mabel Alice*, shortly to be replaced.

'She's been a good servant to us,' said Neil. 'But it's progress. The lifeboats are getting a little bigger and a little faster. We get about 25 shouts a year, summer and winter, about half and half. When the flare goes up, and we all run down, I pick the best men that I have there at the time.'

Andrew advised me not to say flares as they all now used bleepers, but one of the other lads said that they did in fact still use flares in daytime in the summer. I asked how long it took them all to get down.

'At night, about nine minutes,' said Tim.

'In daylight five minutes,' added Patch.

When they went out, did they get assistance from helicopters from Culdrose (HMS *Seahawk*)?

'Yes, we are very well covered down here, as we have an inshore lifeboat as well,' said Neil.

As usual, I asked them about their general beliefs, and one of them put in, 'There are no atheists on a lifeboat,' which I thought was a wonderful remark. As a parting shot I asked them in general what they thought the outcome of all the present troubles with the fishing industry would be. Neil said he thought that it would all come well in the end. I felt that there was hope, with fishermen like Antony Hosking having taken delivery of a new boat the day before.

'I bet he wished he had never seen it last night,' said Andrew. Apparently as he had entered Newlyn harbour he had lost his steering power, and had almost run into another vessel (of which more later!).

I very much wanted to talk to Patch, and also Andrew Pascoe, as I wanted the young person's perspective. They were great friends and had known one another all their lives. The Nora Bolitho Room at the mission was unavailable, so we met at the cottage of my good friends Telford and Sue Strict.

Patch and Andrew had known each other since they were four years old, and were then in their early 30s. Patch was the father of two daughters and a son, and Andrew three boys. They had started school together at Newlyn Primary, then moved on to Tolcarne, and ended their schooldays at Heamoore. I started by asking then the usual thing about the quotas.

Patch and Andrew have played and worked together since they were boys. Here they are working in FH396 with their friend Spencer Robinson. *Patch Harvey collection*

'The way they have done it is stupid,' said Andrew. 'You see, we still catch the fish, the ones that they want to protect, but we have to throw them back.'

They both worked with gill-nets. Patch worked for Harverys & Son, while Andrew and his brother had been partners for about two years. I was very impressed with this information; with Antony Hosking, here were young men presumably prepared to spend thousands of pounds on new boats, while I has previously seen only despondency surrounding the older fishermen. Ron Jenkin had said how in the old days they would go out and catch as many fish as quickly as they could, then get the best price they could, without any of today's worries and all the paperwork.

'I think that really comes down to the fact that we have had it better, but not as good as they did,' said Andrew, 'so I suppose it is even harder for them. We came into it at about the time that it all started, so we've grown up with it.

Patch told me that both he and Andrew had lost quite a few friends at sea, that it was part of what you expected going to sea.

'We went to school with some of the sons of men involved in the Penlee disaster. We knew most of the men by sight, but remember that we were both still at school then. Of course we knew Trevelyan Richards, but then everyone did – he was a legend in his own lifetime.'

Andrew's vessel was the *Lamorna*, a 39-footer. He and his brother caught crayfish, lobsters, ray and mackerel before Christmas, and turbot from May to August. Patch went for crabs, and also netted for pollack. It took me back to the days when I fished off the north pier with my brothers – in those days we caught wrass, garfish, whiting, plaice and mackerel, usually about two hours before high tide. Many times we were chased along the pier by Tony Stevenson for some small thing that we had done not to his liking. If we were over on the south pier we would stand on the roof of the tidal observatory, and Jo Hutchings, who was in charge in those days, would go mad. I was to become great friends with Jo and his wife Ann over the years. Ann was a Quaker, the only one that I have really known; she once took me to the Friends Meeting House in Marazion on one of my many visits home. The simplicity of the place took my breath away. Prior to Ann's death she lived with Jo in a bungalow up in the Gernicks, at Newlyn, and the view from their front lawn over the bay also took your breath away. She was a beautiful woman, and Jo always said to me that she was beautiful on the inside as well as the outside. I am sure that he was right. They both loved Newlyn, as I do.

A view of Newlyn harbour, looking down from the Gernicks towards the Lizard Peninsula. *Michael Charman*

Patch told me that in their younger days he and Andrew worked from the Isles of Scilly for two or three years on a boat called *Alida*; from what I gather, that was the time that the wild oats were sewn. As they say, 'When a juvenile sows his wild oats his parents pray for a crop failure.'

When I asked if they had ever been afraid at sea, they both said many tines. In fact, Patch told me that he went to sea to help out a friend who was short of a hand, only to find himself being rescued by his friend from the RNLI.

I was quite impressed when Patch told me that he had been a professional rugby player with the Pirates for a year. When I asked why he had not stayed longer he said that he became bored and missed the sea. I asked them both if they thought that they would ever leave Newlyn – they both said no. Patch said that he would not even want to live in Penzance. Now, that I can understand.

CORNISH FISH PRODUCERS ORGANISATION LTD

'Our primary function is the management of fish quotas'

It took me about one minute to walk from the Mission to the CFPO offices at 1 The Strand in Newlyn. That is another thing I like about Newlyn – it is so very compact. On arrival I was greeted by two young men, Paul Trebilcock, the chief executive, and his assistant Andrew Wheeler. After a while Paul disappeared upstairs to prepare for a radio interview he was to do later that day. In fact, he was on the television the next evening, and certainly knew what he was talking about (Antony Hosking's new boat).

It seems that a good executive is often one who can delegate all the responsibility, shift all the blame, and appropriate all the credit. Not so Paul. Andrew likewise was a charming and bright young man, setting out on a career that he loved. It struck me that he was a good listener, and the thought came to me that no one ever listens themselves out of a job.

'The CFPO is a company of fishing boat owners,' he explained, 'and our primary function is the management of fish quotas. I like to use the analogy that we run like an official trade union. We act on behalf of the fishermen and anything related to their job. It is a highly regulated industry, as most people probably know.'

I said that I had been talking to a skipper or two, and they had told me about the frightening amount of paperwork they have to do.

'Most fishermen are not into paperwork, they just want to go out and catch fish. They don't want to worry about anything else, but they have to. We help them to understand about many statutory authorities.'

Was Andrew a local man?

'I was born down here, and my father was born in Newlyn, although he was not a fishermen. We moved away when I was two or three, and I moved back down here in my early 20s.'

Was there anywhere he would rather live?

'I don't think so. I have a young family and a job I enjoy, so I see no reason to move.'

TRELAWNEY FISH

'We were thrown in at the deep end'

It was another bitterly cold day when I walked into Trelawney Fish, situated in the Strand in Newlyn. There was still a strong east wind, as there had been all that week. I had gone to see Godfrey Adams, and was fortunate to catch him in, and all the more fortunate to be granted an interview there and then.

'Trelawney Fish are wholesale fish merchants, and have been in business in Newlyn since 1982, when we started wholesaling fish in a very small way. In fact, when we started we didn't have any premises

Newlyn harbour, showing the ice-making plant adjoining the fish market. *Max Channon*

at all. We had a 35cwt van and a bench and a set of scales. There were very few beam trawlers here in 1982; Stevensons were getting into beam trawling, but I am not too sure how long they had been beaming prior to that. A little later, in 1988, we branched out and bought our first trawler.

'Our first premises were a small pilchard store down in the village opposite the fish market. It was about 16 by 16 feet, and a long time before the health regulations we have today. We traded there for about eight to ten years, then looked for bigger premises. We rented some from Newlyn Harbour Commissioners in about 1990, but unfortunately they burned down in 1992. After that we moved back to our old premises and in 1994 moved here, which was part of Shipham's, and previously Cornish Canners, quite a distinctive building for its time. Shipham's fishpastes, canned mackerel and tinned pilchards all came from this site.

'There was a handful of companies here at the time and the industry was quite buoyant, as it was right through the 1980s. The Stevensons extended their fleet, and in the '80s gill-netting took off as well, so we had good fish being caught.

'My partner Jim Tippett had an idea to set up a novel distribution system – a network of people who had vans and we would supply them with fish on three days a week. I had already given up a good job, we were already committed, but it was quite clear that although we had enormous interest from the people in various areas, as soon as we said it was fish the interest waned. We started with one customer.'

He told me that there was one man who had worked this idea for many years and had made a good living from it.

'We were thrown in at the deep end, against some heavy competitors – Suttons, Kismet – they have gone now. We are probably one of the bigger ones left now.'

He went on to tell me that things have changed very much. The usual stories came out about the quotas and all the paperwork that the skippers have to do. As he spoke a slight look of sadness came over his face. When I asked him where his partner was, he told me that he had died of cancer, and you could see that this fact still hurt. But here was another man who was too busy to be miserable.

Godfrey Adams, Managing Director of Trelawney Fish. Having started out in 1982 with a partner and a 35cwt van, they now have a very successful business in Newlyn. *Michael Charman*

ANTONY HOSKING

'I think it is important that it all stays within the family'

Antony is the skipper of a Newlyn vessel and I was very eager he hear his views. In my opinion he is a very brave young man for whom I have great admiration, which I am sure you will share.

Left As a boy of 5 or 6, Antony Hosking used to accompany his father on trips out to sea. He nagged his father until he was given a punt of his own, and by the age of 20 had obtained his first skipper's ticket. *Michael Charman*

Opposite Newlyn harbour in 1980, when the mackerel fleet was at its height. *Antony Hosking collection*

'Really I just followed on with my father,' explained Antony, when asked when he first went to sea. 'It is strange really because when I was very, very young, I never had a thought about the sea. But Dad got me interested and took me down to the quay and showed me the boat. I thought, I like the look of this, and of course back in those days there was a big fleet of inshore mackerel boats. I thought, I fancy this. I could not have been more than five or six years old when I first went to sea with him, on the odd sort of Sunday trip. I used to go with him when they changed over to the nets. They kept the stow pots they used to keep the crayfish and lobsters just outside the harbour. He would take me on a Saturday morning when they used to get their stow pots – at that time I thought that was a full fishing trip.'

This had been a really busy day for him. In fact, we stopped the interview so that we could see a television report made that very morning about the new boat that he and his family had commissioned, and of which he had taken delivery the day before. The *Silver Dawn* – what a leap of faith!

'After a while I wouldn't leave father alone until he bought me a small punt. Then came that long, hard learning curve looking after a boat. Like all things it started very small – I never had much opportunity to do much in a small boat. You see, I always envied some of the other lads who lived in Newlyn and were my age because they could come home from school and jump down into their punt and go away and catch 10 stone of mackerel or whatever was going. But with me living outside a little bit and with Dad at sea all the time, there was no one to take me down to the sea. I could go at weekends. But not being able to get to my boat never put me off. The moment I was able to leave school I did so and started fishing with my father.'

He sounded very keen about it all.

'Yes, I was and I still am. Father had a couple of boats built at Thomas's yard at Polruan. The first one was called *Boy Antony*, named after me. She was 32 feet, and was launched when I was something like five or six years old. Then he had another, but I can't tell you exactly when that was – a 40-footer. Then, when I was 16 and looking to leave school, we looked at buying a second-hand boat, a 60-footer, which we have stayed with till now. The new vessel is a family boat. I don't think any one man could afford a boat like this. I think if he had that money he wouldn't want to go to sea anyway!'

Left The first berth for Clive Hosking, Antony's father, was aboard the ***Renovelle***, a fishing boat family-owned by Jimmy and Joe Madron, Clive's uncles. *Antony Hosking collection*

Below Antony, aged 4, aboard the first ***Boy Antony*** (1976), and aged 6, with his father Clive in the wheelhouse. *Antony Hosking collection*

Antony, aged 12, launching *Boy Anthony 2* in 1984 at Thomas's Yard at Polruan, and in the wheelhouse. *Antony Hosking collection*

When had he stopped fishing with his father?

'We were not fishing together very long. Father was taken ill just after I left school. We didn't do more than a couple of trips together. We'd sold the second *Boy Antony*, the 40-footer, or we were in the midst of trying to sell it and we were working the 60-footer. She was second-hand when we bought her – she was a French boat, an ex-tuna boat and a trawler. We were up the Bristol Channel and we were just hauling in our gear. The next thing Dad was laying face down on the deck. He had had a heart attack. Luckily for us we had on board a fully experienced crew. I was only 16. There was no way that someone of my calibre could jump up to the wheelhouse and take charge.

'He was only in his late 40s. He was airlifted to Truro, I went also. Within a week or so he was home. They confirmed that he had had a heart attack – they also discovered that he was a diabetic. It was a hard blow to me. Me and Dad have always got on well together, we still do. That is why we are here today with the new boat, because we have always stuck together. We support each other. I think it is important that it all stays within the family.

'A new skipper by the name of Frank Dallimore came on to the *Boy Antony* and ran her for a good few years. He took charge and I worked on the deck and helped take care of the boat while he was there. I took control of the engine room and worked my way up, then a few years ago I decided to go to Plymouth and do my skipper's ticket. I took my first skipper's ticket in 1992 when I was 20 years old, and my second ticket in 1995 when I was 23. It was a big step forward for me, because doing something like that gives you the confidence to step up into the wheelhouse, it gives you something to fall back on. That's what it's all about.'

It struck me that you get two sorts of seamen in Newlyn. There are those like Antony, and there are the other sort who can't wait to get into the Swordfish or the Dolphin and spend their hard-earned money.

Right **Antony (left) during the re-engining of *Boy Antony 3* in Penzance harbour.** *Antony Hosking collection*

Below right **Landing fish from the 1988 boat *Boy Antony 3* in March 1997.** *Antony Hosking collection*

'Fishing is a funny old game,' replied Antony. 'Trouble is with fishing, you know the men have had a hard time, it's a hard job. Some work hard and play hard – that's up to them. Maybe living outside Newlyn has made it a bit different for me. You see, I've always had to drive home. Their home might be just up the street. Anyway, I like to come home and spend quality time with my fiancé and family. As it is, I spend a lot of time at sea. When I get the time I like to relax.'

So what had happened to the new boat yesterday?

'We bought the vessel over and did some trials with her. She went up on to the slip at Polruan to have some other work done on her. Admittedly we'd not been out in any poor weather. Yesterday, with a south-easterly 8/9 forecast, we left Falmouth. I'm pleased to say everything went very, very well, and once we got round the Lizard Point we had fair weather towards Newlyn, and I was delighted. But a south-easterly is the worst wind you can get into Newlyn through the right gaps. It can make it awkward. I decided we could make a run for it, to get in. It was just prior to this that I had been down into the engine room and into the steering compartment to make sure that everything was OK. We came through the gaps to give the men time to get the ropes and fenders ready. The weather was that bad outside now that I couldn't let them go out on top of the shelter deck.

'I nudged her astern just to stop her dead in the water – naturally the wind was carrying her towards the slipway. I decided to give her 8-10 degrees starboard helm in gear to bring her head around a bit. My worst nightmare was coming true – the helm indicator didn't move and it started up 'No rudder response' and this alarm started. This boat was new to me, and it was new to all of us. Although

Antony's latest fishing boat, *Silver Dawn*, built in 2002 at Falmouth and photographed in Newlyn harbour in 2003.
Antony Hosking collection

you've done training, it just could not have happened at a worse time. That's not true – it could have happened just outside the gaps. There is no doubt about that – if it had happened half an hour earlier we would have been coming around the Lizard. We could have rigged up the emergency steering – it only takes 10 minutes to rig. Just think, we could have gone aground in the harbour somewhere. I tried calling Newlyn harbour and anyone who was around, but no response – it was dinnertime and there was just nobody there.

'I saw the small boat that moves Stevenson's beam trawlers around. I was waving to the chap on the *Chickadee*, but he didn't see me – I was screaming and screaming. There was a big reception – 10, 15, 25 people on the quay, all waiting to see the new boat come in. Family and friends, and other people who had been involved with the boat. They were all thinking to themselves, what the hell is he doing over there!

'When you come about boats usually square themselves up into the wind. I thought, I hope she does. Sure enough she squared herself up. With the rudder locked hard to port, I kept coming astern. I could see us going sideways and we were going towards the south pier, but she was squared up to the quay and I thought, if I could keep nudging her astern… Judging by the wind she was going to run in along the side of the quay parallel, and she came right aside the quay. The boys could see what I was attempting to do and they managed to get some fenders out quickly. There was another thing you would never expect in that sort of weather. There were a couple of fishermen on the quay – they had gear there. With the sea crashing over the lighthouse, I expect they were worried it was all going to get washed off, so they were lashing their gear down, which was a perfect opportunity to chuck them a couple of ropes. I thought to myself that there was someone up there looking after me. It was not the first time in my career that I had thought that.'

What did he think of all the present-day fishing restraints? How was it affecting him financially?

He looked rather sad, and, with his head down, replied, 'Yes, it is affecting everybody. The problem is that you have got to have conservation. If managed correctly there would never be a problem. I am not saying that how you go about it is easy, because you are always going to upset somebody. What you have got to recognise is that there are so many methods of fishing. When the general public hear about fishing it is always a trawler and the word trawler is used in general terms for the whole industry.'

As I had asked other young men in the industry, I asked Antony if he considered that he was earning a decent living to support his family in the manner that he had been used to. They had said yes.

'Please don't get me wrong,' replied Antony. 'I earn a decent living. What I am trying to say is that we are all earning a decent living, but the trouble is the investment you have to keep putting in year and year to get that living. You see, you have so many tons as a fish quota, and they keep cutting it in the belief that those fish are getting scarce – when it really hurts and digs in is when they have cut it by 50 per cent. So we say to ourselves, OK, we will go and target something else, we will avoid what we cannot catch. The trouble is that the ones you are trying to avoid seem to be thicker and more of them – bigger as well as thicker. You see then that the authorities are wrong.

'Our own British Navy come on board and talk to us. They will admit that they are only the police of the sea for the fishing industry; they are on fishing patrol and they have a job to do, and they do it well. They can see it is not right. If you chuck over the side 2 tons of dead cod, what's the point? There have got to be better ways of conservation within the regulations. There is no more fiddling going on. I land my fish in France at Guilvenic. Four hours before I get into the French harbour I have to notify the fishing authorities by fax. We have to tell them exactly what fish we have on board. Naturally, when we get there the inspector comes down and checks, and also when I come home I have to leave part of my logbook in France, which has the record of what we have landed. I have to bring my landing declarations home and present them to DEFRA, and they check them as well.'

I told him that I admired him tremendously, and thought him terribly brave. If we have people like him in the fishing industry, have we anything to worry about...?

ELIZABETH STEVENSON

'I don't think that there is a bigger privately owned fishing company...'

Elizabeth Stevenson. *Michael Charman*

It must be said that without the Stevensons' story this book would be incomplete. If it were not for the Stevensons there would be many out of work in the port. In fact, since the 1700s the Stevenson family has been a major force in Newlyn, as Elizabeth explained:

'I am the great-great-granddaughter of William Sampson Stevenson, a Newlyn boat-owner and fisherman. He married locally and had a son called William. I am also the great-great-granddaughter of James Richard Harvey, a fish merchant from Great Yarmouth who came on a fish-buying visit to Newlyn, married locally and had a daughter called Sara. In 1889 William married Sara and joined his mother-in-law in business trading as Stevenson & Harvey. William and Sara had many children, and in the 1920s four of their sons joined their father in setting up a fish-selling agency, a grocery shop and a bakery, trading as W. Stevenson & Sons. So that's where the original partnership comes from, in the late 1800s and early 1900s. But in fact the business of fish merchants and boat-owners was way back in the 1700s – that was the real birth of W. Stevenson & Sons.

'The partners have changed over the years, but W. Stevenson & Sons still exists and probably has one of the largest privately owned fleets in Europe. I don't think that there is a bigger privately owned fishing company with more vessels than we have. We have 35 vessels currently, all fishing out of Newlyn. We have 24 beam trawlers, a couple of side trawlers, a couple of netters, all working in the Western Approaches, and some under-10-metre boats that work in and about the port of Newlyn. We land approximately £10 million worth of fresh fish from these vessels each year. Each time we land some, it is made up of 53 species of fish. Most are the high-quality prime species, and they are in demand in the European fish markets of Spain, France and Italy. A huge percentage of fish landed here in Newlyn is exported. We have some 120 seagoing men at any one time, and we have in excess of 60 employees on shore looking after our vessels and fish auction, and our fish marketing on a daily basis.'

Elizabeth's whole interest is in the

Some of Stevensons' boats: ***Billy Rooney*** (PZ532, 31.86m, built 1973); ***Marie Claire*** (PZ295, 26.22m, built 1946); ***Basil St Clair Stevenson*** (PZ99, 26.12m, built 1969); and ***Lisa Jacqueline Stevenson*** (PZ476, 24.20m, built 1973). *All Dave Willacy collection*

Above A 'WS&S' trawler being overhauled in Newlyn harbour. *Max Channon*

Below Skipper Dave Willacy repairs nets aboard the Stevenson boat ***Filadelfia*** (PZ542). *Dave Willacy collection*

Stevenson company and Newlyn, but what of the fishermen who are very worried about quotas? What did she think the future held for the industry, at least in the short term?

'Fishing does mean an awful lot to me. I have been brought up with it all my life. My family are involved with the business, as is my husband, who is also a boat-owner and a fisherman. I am very active in the industry and love anything to do with fishing. But I am very concerned about the industry in the short term. We have got into this state mostly by mismanagement, whether by DEFRA or the Government, and from the things that come from Europe and the way our country interprets then.

'I think that we have got a very sound future in as much as there will be a good fishing industry here in ten years' time. I don't think that we have reached the bottom of the trough yet, and I think we are still waiting for that turn-around. The great strength that we have against some of the other communities around the country that are dependent on fishing is that we have that huge variety of species here in our waters, and having that variety means that we are not dependent on any one, two or three species. It's a huge mixed catch. I think that is our safety net – unlike the North Sea, where some of the species, predominantly cod, haddock and whiting, have a serious problem. We don't have that problem, and in fact we do have more fish here at the moment in 2003 than we have seen for many years. We can't understand why the scientists' evidence doesn't back that up, because what we see is what we catch and what we land on a day-to-day basis. The scientists disagree with that and we are fighting frantically to try to come to terms with the way the scientists get their evidence.

'So we are doing a lot of investigative work ourselves and we hope that we can prove that the way that the scientists do their catch assessments is flawed and that their statistics and data are wrong. We are trying to prove it. Once we get the correct evidence we think that we can prove that stocks in the South West are not in decline, as is thought by the scientists. Therefore we can increase some of the quotas to a level that some of our boats will not be restricted. Then it will be much easier to operate. This could happen within a few years, especially if we introduce some no-take zones and closed areas as well. We believe it will be here in 2013. It might not be the same industry as we have here now, but I am pretty certain, and with God's help, that there will be an industry.'

MICK MAHON

'I have yet to see anything come out of Brussels that has conserved one fish'

'I was 17 when I first went to sea,' said Mick. 'I was a "service brat". My father was in the Army, and I lived all over the world. I really wanted to go to university but I had asthma and was very ill during my GCEs. I just walked out of school one day and thought, I'm going to be a fisherman. I went labouring for about a year to build myself up, trotted up to Grimsby and ended up fishing out of there for about 12 years, then another four or five years working in Iceland, Norway and Greenland, plus the North Sea. I wasn't happy working out of Grimsby – it was very strongly union-controlled. The cost of landing fish there was astronomical, and I was also a bit worried that if one of the North Sea rigs blew it could wipe out the fish in the North Sea. So for various reasons I came down to Cornwall. I could see there was future in the fishing industry here. To a certain extent there was. I've watched a little bit of expansion in the fishing fleet in Newlyn. But a decline also, with all these rules from the EC.'

I asked him about the boat he decommissioned some time ago.

'I decommissioned a vessel about ten years

Mick Mahon is a fisherman of the old school, and is very politically active. He provides insights into the decommissioning fiasco, and expresses his views on how to halt the decline in Britain's fishing industry. *Michael Charman*

ago. I put in for decommissioning, but to be perfectly honest I was not too much interested. I had a good little boat, it was working reasonably well. The boat was worth about £35,000, I put in for £50,000 for decommissioning and was shocked and stunned when I was offered it. I couldn't afford to turn it down. I got the money and scrapped the boat. Most people who had wooden boats employed someone to scrap them, but in my case the local scrapyard agreed to cut it up and they would have all the metal out of the boat for scrap. So it didn't cost me anything to cut the boat up – that was one less cost that I had. Some owners of boats had to pay up to two, three, four thousand pounds, depending on the size of the boat that they had to cut up.

'I got a smaller boat without a bank loan around my neck and cleared all the debts off the previous boat. So it was a fresh start for me. For the last two years I'd owned the *Bett-G*, a 40-footer, I had worked single-handed. People say it's hard work. I say, look, I don't drink. I can't drink. My liver has packed in, I'm a diabetic, and I've had a heart attack. I've been working single-handed for the last ten years. I'm a trawlerman. It's my own boat now, the *J-Anne*, a 29-footer. I don't have any crew. I used to do a 20-hour day, but as I've got older I've cut it down to 16 hours. You work at your own pace. You don't have to worry about anyone else.

'I don't like the idea of decommissioning, but I couldn't afford to turn it down. They were giving me more than the boat was worth. On top of that I salvaged a lot of parts off the boat – fishing gear, bits of engine, electrics – which I was able to use on the new boat and save myself further costs. The idea of decommissioning is to get rid of the British fishing fleet – reduce the fishing fleet so there's more room in our waters for our European partners. It's well known that 80 per cent of north-west European fish are in our waters. We British-registered boats are only allowed about 12 per cent of the value of the fish in our waters, and that 12 per cent also includes the flag-of-convenience vessels. You take that into account, and genuine British boats are only allowed 8 per cent of the value of the fish in British waters.

The ideal example is off the coast of Cornwall. Here we are allowed 7 per cent of the cod, while the French are allowed 78 per cent. Off the west coast of Scotland, in the deep water, just within the last few months British fishermen are allowed 2 per cent of deep-water species. It's the same all around the coast of Britain. British fishermen have a far smaller share of fish than our EU partners – with ten more countries just joining.

'I think that now, with the latest government measures coming in, it's a ploy to reduce the British fishermen again to accommodate the newcomers. Let's put them around the coast of Britain, let's get rid of some more British fishermen and then there will be room for them.'

What action should be taken?

'I wonder if it's not too late. I personally

'The idea of decommissioning is to get rid of the British fishing fleet,' says Mick, 'to reduce the fishing fleet so there's more room in our waters for our European partners.' These dramatic photographs pose the question, is this the end for the British fishing industry? *Max Channon*

would like to see a lot of action taken. I have, in fact, taken action myself in the past, as have other fishermen. Our problem in this country is that there are about 18,000 fishermen, and that is in the whole of the UK. We have got dozens and dozens of organisations – we are divided, which suits the government. Because we are divided, they can rule us. If we were to unite, we might have some clout – we are not a lot of people, but we have a lot of sympathy from the public. Politicians don't take too much notice of us as there are only 18,000 of us.'

Mick stood up and walked around a little, stretching that huge frame. He did indeed look the part with his big bushy beard. I liked his enthusiasm, his energy, his devotion to his cause. He sat down again and went on.

'Not too many votes, you may think! But if you work it out there are between 30 and 40 constituencies where fishermen can have their say, where they vote. Take their wives into consideration, their aunts, their uncles, the people who work ashore, the filleters, the fish processors, the engineers – we start thinking then that the opinions of the fishing industry could actually change who the local MP is going to be. We need a united voice to do that – unfortunately we haven't had that. The only time we have ever been united in my fishing career was in 1995 when the whole of the British fleet flew the Canadian flag.

'I had just finished watching *The World At War* on television, the series about the Second World War. It had originally been shown 20 years before. It showed some Canadian troops going ashore at D-Day with the British troops.

'This thing blew up on the Canadian east coast when the Canadians arrested a Spanish boat called the *Estai*. Our government supported a fellow member of the EU – Spain – rather than Canada. I was a little bit upset over this – I thought we owed Canada a great debt. They had fought with us over two World Wars, they came and fought for their mother country, Britain. I thought that was typical of how Britain stabbed its friends in the back. We have a record for doing this. I faxed a letter of support to the Canadian High Commission. My stepdaughter was living in Canada at the time. We rang her up and asked if she could send a Canadian flag over. In the meantime another fisherman told me he had rung the Canadian High Commission for 200 flags. He wanted me to contact them again for him as he didn't have a phone. He gave me the contact number, and I rang for him. In the meantime I trotted up to Yeovil and made a speech at a rally. Speaking on the same platform was Norman Tebbitt, Bill Cash, and Christopher Booker. There were about 1,000 people there. I said that as a fisherman I was ashamed of my government with its support for Spain. The place was in an uproar. They were cheering and yelling – it rather surprised me as I would describe them as a blue-rinsed Tory set! I told them how we had contacted the Canadian High Commission to see if they had any flags. Afterwards I had dozens of people coming up to me to give their support.

'My wife had some old newspapers she had brought back the previous year after a visit to her daughter. I rang a couple of them but they were not very interested. I ended ringing the *Toronto Sun* and talking to a guy called Rob Benzie. I asked him if he could put an appeal in the newspaper for some flags, and he said he would ring me back. After about 10 minutes he rang back and asked if they could bring a few flags over. He told me that he would bring a photographer. I asked him if he wanted any media coverage, but he said no – he wanted a scoop. He was on the quay 36 hours later.

'They had been to the nearest shop and bought their stock of flags and brought them over. They slapped them over a boat called the *Snowdonia*, took a few photos and went back to Canada. It was a reasonably big story over there. The story didn't pick up in Britain till four days later, then it was on the front page of *The Times* and the *Telegraph*. Then it all blew up. The total number of flags that came over from Canada was 100,000. Britain

Mick Mahon with Brian Tobin, former Canadian Fisheries Minister, and Aileen Le Breton, manager the band John Prince and a Piece of the Rock. *Mick Mahon collection*

was ringed with flags – 99 per cent of the British fishing fleet was flying the Canadian flag. They were not only on the boats, but here in the village – we had them on the fish and chip shop, the car park, the Post Office, and so it went on.

'There was another story that was never picked up by the press and it rather disappointed me. In Ireland, north and south, they flew the Canadian maple-leaf flag, not the Union flag or the Irish tricolor. We even had reports of it being flown on Danish vessels.'

I asked him what Norman Tebbitt had said to him.

'He said that he thought it was a good idea, said he hoped that it would take off. We had a bit of a chat, as under Margaret Thatcher he was a Junior Fisheries Minister. He did say that he thought that fishermen worked very hard.'

If he was a Fisheries Minister, what would he do?

'I would say to my Prime Minister that the only chance for the industry to survive is complete withdrawal from the Common Fisheries Policy, and Britain to take control of its own waters again. I would like Britain to withdraw from the EC altogether. I am a member of the UK Independence Party. I used to be a Tory. They are running around now saying they would withdraw from the Common Fisheries Policy. I'm sorry, but we had all those years of Tory government, and all the time they were stabbing my industry in the back. Every year the Fisheries Minister sits down and signs 14 pieces of paper giving permission for the EU countries to fish in

British waters, including those well-known fishing countries Luxemburg and Austria! If he didn't sign that piece of paper, come 1 January no other country would be able to fish in our waters inside the 200-mile limit. We have all the aces but the government never uses them.

Fourteen years ago, when I became politically active, when I had to throw 90 per cent of my catch back into the sea dead, I had the best cod fishing I have ever seen in my life. You could almost walk around the coast of Cornwall with cod. The trouble was that the quotas were caught within a matter of weeks. There was so much cod we had to dump fish. I was dumping up to £6,000/£7,000 a week. I did that for three or four months. In the end I broke the law and I landed it. We were dumping fish when there is starvation in the world. I have yet to see in my fishing career anything to come out of Brussels that has conserved one fish. If anything it has an opposite effect. What it means is that you stay out at sea longer catching fish you are allowed to catch, dumping the cod or whatever the banned species of the month is. So it actually increases your fishing effort. If there is any destruction of the stocks, it is purely down to the mismanagement of our government and Brussels. They're happy if, at the end of the year, the bureaucrats can sit across the desks from each other and push little paper fish across to each other to match up. And that is what it is all about.'

FISHERMAN'S GRAVE

Did he land all his fish at Newlyn?

'I land it all in Newlyn. I have to be in harbour each night. All my fish goes through Stevenson's.'

I felt that someone with such commitment as Mick could do almost anything he wanted to. He smiled and shook his head.

'You might be right. Do you know, that thing with the Canadian flag, the Government was supporting Spain. In the end, with the support we had, the Government did a u-turn. They stopped backing Spain and started supporting Canada. Not only that, but we pressurised the Canadian government as well. The Canadians have always been quiet on the political stage, they have always kept a low profile. With all the pressure they were getting from Brussels they were all set to give in. They decided to stand firm.'

'You started that.'

'Yes and no – another fisherman came up with the idea of flying the Canadian flags. In fact, the person who did the most work was my wife – we were getting 300 phone calls here a day. At 9pm she had to pull the plug out of the phone to get some rest.'

They say that there are three steps to accomplishment: plan purposefully, proceed positively, pursue persistently. I was sure that Mick was well enough equipped to do all of these.

Part 2
The Scottish Fishing Industry

The Birth of Fishing on the Moray Firth

Sandy Sutherland

No one knows exactly when fishing began along this part of the Moray Firth coast. We do know, however, from early records that fishing was taking place from the various creeks and bays that could offer a measure of shelter to the boats, and that is why so many of our coastal villages are to be found off the beaten track, so to speak.

It is known that the first house in Portknockie was built in 1677, and the village grew relatively quickly from then on, so that by 1792 there were 80 houses, and its menfolk operated 16 fishing boats. Findochty, which had came into being around the same time, was smaller, with 45 houses, 10 boats and a population of 162. In 1727 Hay, the laird of Rannes, had five houses built at Portessie for the fishermen he attracted from Findhorn and also Maviston, a hamlet further west. By 1792 the village had 44 houses and 12 boats. All of the fishermen in these places were engaged in line-fishing, mostly for cod and haddock. In 1791 Buckie had 102 houses west of the Buckie Burn, and 63 houses on the east side. All of the fishermen were engaged in line-fishing, mostly for cod and haddock. Just as in the case of Portessie, the superior or landlord played a major part in the establishment of the fishing stations. In the case of Portknockie and Cullen it was the Earl of Seafield, with Ord being largely responsible for Findochty. Buckie consisted of two parts: the lands of Easter Buckie were in the hands of Gordon of Cluny, while the Duke of Gordon and his 'tacksman' (lessee), Dunbar, held dominion over the lands of Nether Buckie, later to become Buckpool.

During the 18th century a cluster of nine houses was built above the cliffs in what is now known as the Seatown of Buckie. They were referred to as the 'Nine Holes' or the 'Nine Os', since outside each house the fishermen dug a deep hole. Mussels were the bait used in line-fishing, and when the men were at sea the women used to go down to the 'scaup' (mussel bed) on the sea shore, fill a basket with mussels and carry them home. When the fishermen returned from the sea with their catch, the lines were taken home and made ready for an early start the next day. The women shelled the mussels and baited the lines before tossing the empty shells into the hole. Of course, at this time there was no sanitation so one can imagine what else landed up in the holes.

Above Portknockie harbour full of sailing boats at the about the turn of the 20th century. *Buckie District Fishing Heritage Museum*

Below The Findochty herring steam drifter *Bon Chief* (BCK17). *Buckie District Fishing Heritage Museum*

In 1770 cod fishing off the Caithness coast failed. The merchants from Dunbar encouraged the fishermen to fish for herring, which were abundant near the shore; they were only too glad to take this advice, especially since the government was offering a bounty of 50 shillings a ton for herring. The season proved to be most profitable and continued to be so; by 1780, ten years later, there was something like 100 boats engaged in herring fishing out of Wick.

In 1819 a man called J. F. Donavon, a fish buyer from Leith, hired six experienced Dutch curers, and the quality of their cure became the standard adopted by the Fishery Board. This was called the 'Scotch Cure', and to receive the 'Crown Brand' herrings had to be gutted, salted and packed within 24 hours of being caught. The Crown Brand was the seal of quality and was recognised as such throughout Europe.

In 1786 a number of influential Scotsmen, led by the Duke of Argyle, set up the British Society to extend the fisheries and to improve the sea coast of Scotland. Land was bought, piers were built and later harbours. The first pier at Buckie was built at this time at the mouth of the Buckie Burn, on the Buckpool side at Bridgend. Built of stone and wood, it was used for the loading and unloading of vessels. Unfortunately it did not stand long before it fell to the power of wind and wave.

The Napoleonic Wars had a disastrous effect on the herring fishing as the European market was cut off. Trade, however, was established with Ireland, the West Indies and the Americas. Slaves on the plantations had to be fed, and herring was both cheap and nutritious. When the wars came to an end, trade links were quickly resumed with the countries of Europe, but such was the demand for herring that markets could not be satisfied. During the summer months daily cargoes of herring were shipped from local outlets to the Baltic ports of Stettin, Königsberg and Danzig. At this time gutting and curing stations were set up at Portsoy, Cullen, Findochty and Buckie.

A fisherman's wife looking out to sea hoping to see her husband's boat on the way in. On her back is a 'creel', which would be used by the womenfolk to take fish from the boat to barter in the local countryside for eggs, cheese, butter, etc, which they would then sell in the town. *Buckie District Fishing Heritage Museum*

Cork was used to support the headlines of the drift nets used to catch the herring. The nets hung just below the surface, but where they were joined together something more buoyant was needed, so the fishermen bred cats and dogs. When the animals had grown to a certain size they were killed and skinned. The skins were sewn up, dipped in tar, dried, then blown up through a pin inserted in one of the hind legs.

At this time Portessie had 84 boats and 268 fishermen, and Buckie had 250 boats with 900 fishermen and boys. Curing stations and ships' chandlery stores sprung up everywhere. In Findochty alone there were four curing firms. Shops of all kinds – butchers, bakers, tailors, grocers, boot and shoemakers – came into being.

Portknockie, ______ 190

Alexr Mair Rose & Sons

586 Beautiful Star

To John Slater,

ROPE AND SAILMAKER.

1901 Statement

Date		Item	£	s	d
April	3	Fore Sail 300 yd 2 cotton Miz 200 yd 5 cotton Jib 150 yd 12 cotton	53	10	–
		Small Miz 61 yds 3 cotton Drag bag finished bails & etc Capstan Cover	6	10	9
		2-15 in triple blocks one wire & 1 iron stroped 2-11 in ditto iron patent	7	6	9
		2 Single 7 in Pat button blocks & pinnets		14	6
		1 Double 1 Single 8 in 1 Double 7 & 1 Single 7 in purches	2	6	9
		2 Single 6 in button mizzen wire pinnets & hooks	–	13	3
		2 Double 11 in comm long strop 2 Double 8 Miz Sheet	3	4	3
		2 Single 8 in & bullseys fore sheet metal boom block & chain	2	10	1
		fore tee jib tee Miz tee miz stay with hook & etc	5	13	9
		fore & Mizzen parley rollers & small miz Parley L.V. rollers	1	3	6
		Life buoy 2 cork fenders 2 buckets 2 brooms & handles	1	15	3
		50 fms 7 in coir tow rope 28 lb spunyarn 6 3/4 Galv snatch block	7	2	6
		Side Lamps 1 anchor lamp 1 Dropatic lamp Cabin lamp	4	5	6
		Torch & fog horn 2-10 gallon tanks 1-8 inch board of trade bell	2	4	3
		1 Eight Day Clock 1 Barometer spirit compass & 1 bose compass	4	17	6
		2 Black anchors	3	10	6
		2 herring scoops & jib traveller	–	13	–
		Drag square 2 tugle & bogie L.V. rollers	5	5	–
		Side & riding Lamp stands & boards	2	10	6
		set Galv safety rails & drinking can	3	–	–
		1 & 3/8 Galv 30 fms chain	8	8	3
		all running Gear best Yt manilla	16	18	6
			144	4	4
April	4	By Cash	50	10	7½
		Bal	93	13	8½
Sept	14	By Cash	10		
		Bal	83	13	8½
	16	Galv clip hook & thimble Steel wire fore tee & hook	2	4	8
1902 April	22	112 lb Yt manilla stoppers cork fender hank houseline ball twine		17	1½
			86	15	6
1902 Nov	29	By Cash	10		
		Balance	76	15	6

Carr Forward

During the 19th century ships' chandlery stores and boat suppliers sprang up everywhere. Here is a sizeable statement from a Portknockie rope and sailmaker for 1901/2.

The steam herring drifter *Winner* (BF1805) leaving harbour. *Buckie District Fishing Heritage Museum*

Above Finishing the construction of the north pier at Buckie in the early 20th century.
Buckie District Fishing Heritage Museum

Below A bill for coal from a Wick coal merchant in 1925.

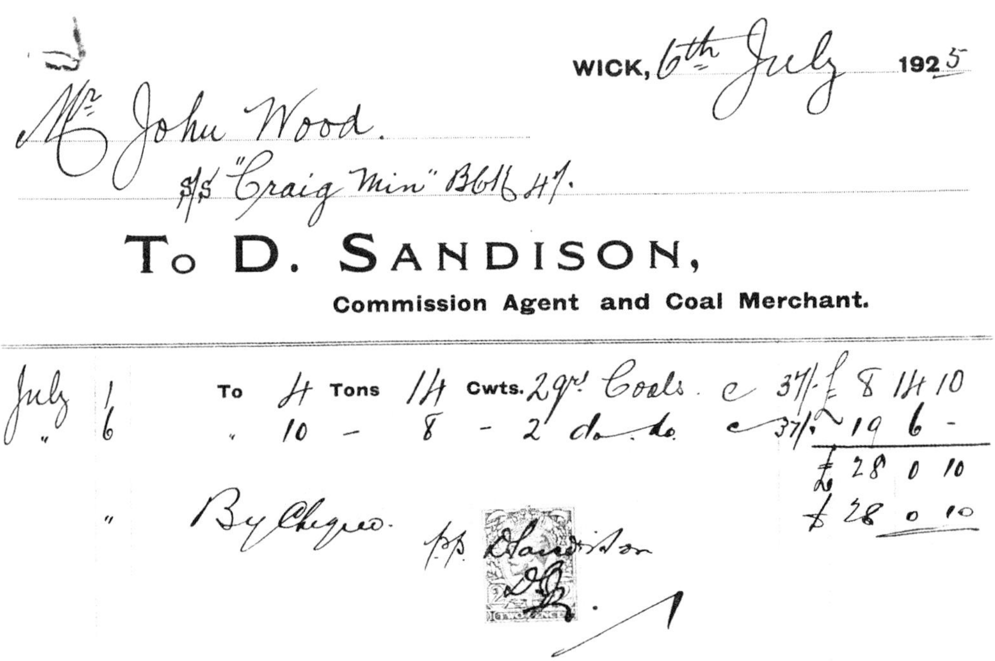

WICK, 6th July 1925

Mr John Wood.

S/S "Craig Min" BCK 41.

TO D. SANDISON,

Commission Agent and Coal Merchant.

July	6	To 4 Tons 14 Cwts. 2 qrs Coals @ 37/-	£ 8 14 10
"	6	" 10 - 8 - 2 do. do. @ 37/-	19 6 -
			£ 28 0 10
"		By Cheque	£ 28 0 10

p.p. D. Sandison
D.G.

At the turn of the 20th century a new vessel came into the arena. Built solely for drift-net fishing and powered by steam, this vessel was to revolutionise the herring industry and indeed change the faces of the communities along the coast. Known as a 'drifter', she had a speed of 8 to 10 knots, and was the ideal vessel to follow the herring shoals from Shetland in May southwards via the ports of North East England to East Anglia in October and November. In 1913 Buckie had 235 drifters, each with a crew of 9-10, the largest fleet of any port in Scotland.

It has been written that for every fisherman going to sea at least six people ashore were directly involved in the industry. The coal-burning drifters spawned large coal merchant businesses, some with their own cargo boats, as in the case of McWilliam of Buckie. Large curing stations were required to cope with the massive amount of herring being landed, while the shipyards grew in size to cater for the construction of the drifters, which measured up to 90 feet in length. Harbours also increased in size to cope with the large drifter fleets, Cluny Harbour, Buckie, being only one example. There can be little doubt that with the drifter came prosperity; whole streets of new houses were built in each of the fishing villages and towns along the coast, still occupied today. This period is referred to in fishing circles as the era of the drifter.

In 1914 came war with Germany. Most of the drifters were commandeered by the Admiralty and together with their crews were forced to forsake the fishing and go off to fight a war not of their making.

After the war, when the majority of the boats returned, the fishing was good for a number of years, the fish stocks having had time to recover over the four years of the war, but the late 1920s saw a fairly dramatic decline. Other problems faced the fishing industry: the price of coal had risen steeply while Britain had lost some of its European markets and, worse still, the country was heading towards the Depression. During the 1930s the drifter, which had brought so much prosperity to the fishing communities, now brought ruin to many families. Some of the steel ones were sold for scrap, but many of the wooden ones were driven up on to the beaches where they were cut up for firewood. Worse still, many a family had to bond their house to the bank in order to get money to buy food. In fishing circles this era is referred to as the 'hungry Thirties'.

In 1939 war with Germany came once more, an event that inevitably led, in time, to further changes in the fishing industry. The changes since that time have been many, and are as convoluted as anything that has gone before, and as always have had a significant effect on the communities, the people and the fishing industry in this area.

Aspects of the Scottish fishing industry

George Behrend

Most people know that, after the Second World War, Russians scooped up large quantities of fish, and now stocks are so depleted that European Union Fisheries Directives are put out to stop Scottish fishermen fishing. Otherwise some species will become extinct, notably turbot and cod, while herring has become a luxury. The traditional waters around Scotland have been worst affected; many Spanish fishing boats, having scooped up all the fish in Spanish waters, came up to Scotland in the 1990s, but now seem largely to have departed.

Fishing is a rough, tough industry, and the rewards were very low for the crews, owing to the structure of the industry with, in most cases, at least one middleman between the boat-owner/skipper, his share fishermen crew, and the consumer. Rewards are now higher, but the restrictions make the payment of running costs, such as being hauled ashore on slipways for dry-dock repairs, etc, difficult to meet; many fishing boats have been decommissioned, reducing the size of the fleet and providing fewer jobs.

Many north-east Scottish fishermen make more money servicing the offshore oil platforms, whose staff move around by helicopter, but which need servicing by fishing boat for almost every other item. The large fishing port of Peterhead is the most easterly port on the Scottish coast. Dr Beeching and Mr Serpell tore up its local railways, without much bothering to visit or consult the locals; in the last century, especially between 1900 and 1960 or so, these were the lifeline of the industry to get the catches sent south to the UK's industrial towns and cities. Indeed, all railways in north-east Scotland have been destroyed north of Aberdeen except the 110-mile line to Inverness via Inverurie, Keith, Elgin and Forres, inland all the way except between Nairn and Inverness. Thus the Peterhead fish catch, for instance, all goes by road, apart from a small amount that goes to Russia by sea.

The 19th century saw gas lighting and cooking with coal gas, and coal-fired steam fishing boats were more common than diesel-powered types, unlike today, and they were also lit by gas. The present generation takes mobile telephones, personal computers, television and radio, and mains electric light for granted, but none of these things existed universally 50 years ago, certainly in country districts, although cities had long had electricity, used for trams (ruthlessly replaced by motor buses) and for a few railways from the late 19th or early 20th centuries. Battery-powered radio advanced rapidly in the 1930s, with radar available by about 1939, although electricity was not available for everybody until about 1960. At about the same date, oil began to be discovered in the North Sea, instead of needing to be imported, and steam gave way to diesel at sea, totally, as regards new vessels. While in the past ice houses and factories flourished around all fishing ports of suitable size, today the miniaturisation of all sorts of electrical and electronic equipment, to make it light enough and small enough for use in the space industry, has enabled enormous strides in the development of refrigeration, especially refrigerated road transport, while fish retailing has switched from small independent fishmongers to supermarkets.

In Britain 'fish and chips' has long been the national dish, especially in the Second World War, when other food was short and rationed. Formerly the fish concerned was herring, usually cod, while nowadays haddock is frequently found in these frying establishments. The British public were very conservative in their eating habits before the Second World War, and slow to discover new tastes, being happy to retain the old ones after rationing ceased to disrupt them. Between 1918 and 1939, and from 1945 to about 1973, hardly any fish was exported to the EU because it was fiercely resisted by the EU's original six members, such as France and Belgium. A 1918-39 exception was crayfish, a French delicacy never eaten in Britain then. Now French refrigerated lorries collect at Cornish ports, and the Cornish sail into French ports, and Scottish fish goes direct to France and Spain, as well as to England, by refrigerated lorry. Germany did not import any fish from Iceland and the Faroe Islands as the rights belonged to Denmark. Before and after the Second World War herring was exported to Germany iced in boxes from Great Yarmouth and Lowestoft to Altona and Bremerhaven in boats called 'freshers'. The herring were also rough-packed with salt in barrels. This process was called 'rousing'. The herring were not gutted, unlike cured herring, which were gutted and heavily salted. In 1947 boats would go over to Holland with fresh herring that had been sold in Lowestoft and bought by Dutch buyers in bulk. A sample of the catch was all that went ashore at Lowestoft.

The situation was very different between the World Wars, when fish was relatively cheap both in France and Britain, and there was no European Union. Germany had many financial problems, being forced to pay reparation for the Great War. After the second World War, Scotland, Norway, Germany and Belgium all suffered very severely due to over-fishing by a large fleet of factory ships. These were mainly large and small trawlers of foreign origin, and they used curtain-sized mesh to catch herring in the North Sea and in the process almost cleaned up all the white fish stock. This was very apparent in the Norwegian and Fladen grounds, and it was 1980 before the white fish stocks returned to normal.

In addition, the traditional movements of fish shoals started to change after the Second World War. Every year since about 1880, or maybe earlier, the Scottish fishing fleet had come south to Suffolk and Norfolk, when there were great shoals of herring. They brought their wives in their boats, who lived ashore and gutted the fish, which then went to London, more quickly and cheaply than from Scotland. Later on the wives came by special train to Lowestoft and Yarmouth from Aberdeen, so as not to waste any ship's time getting to the fishing grounds. This was organised by the Scottish fish curers, who were the wealthy element in the Scottish fish business. In the Buckie Drifter Museum there are photos of these women, from Findochty, Cullen and Buckie itself, taken at Lowestoft. The East Anglian boats, like the Scottish fishing fleet, also went to Shetland at the right season. But most people in Lowestoft had only a hazy idea of where these places were in Scotland, being some 200 miles north of Edinburgh. In 1913 the British port with the greatest number of registered fishing boats was Lowestoft ('LT'), and the second was Buckie ('BCK'), with 698, of which 276 were steam drifters out of a total of 884 throughout

Above **Steam drifters returning to Lowestoft from the fishing grounds.** *Buckie District Fishing Heritage Museum*

Below **Gutters curing herring before placing them in barrels.** *Buckie District Fishing Heritage Museum*

Scotland. There were only 624 steam drifters south of the border (presumably including Welsh ones).

Mention was made earlier of the importance of railways to the fishing industry. The railways were put under severe strain during the First World War, so they were 'grouped' into four large concerns in 1921, efforts to nationalise them being put off until the end of the Second World War. Coal was then also not nationalised, and was very cheap, despite the General Strike of 1926. So the railways built new steam locomotives to go faster than the old ones. The London & North Eastern Railway (LNER) led the way, with famous locomotives like *Flying Scotsman*. But its designer, Sir Nigel Gresley, also designed new engines specially for fish trains. Building entirely new vans to match was too expensive, so only some new insulated vans with 'Fish' written on the outside were built, with screw couplings and vacuum brakes but no refrigeration; the insulation prevented the sun and warm air outside from melting the ice placed in and round the fish boxes.

Although there is not enough room here for railway history, what is essential to the understanding of the commercial fishing story at Buckie is that the Great North of Scotland Railway (GNSR – part of the LNER from 1921) was the arch-enemy of the Caledonian Railway (part of the London Midland & Scottish Railway group – LMS – after 1921) and of the Highland Railway (also LMS), which hemmed it in. The North British Railway (LNER) reached Aberdeen (and the GNSR) only by means of 'running powers' over 38 miles of the Caledonian from Kinnaber Junction (famous as the finishing point of the 'railway races to the north' from London to Aberdeen in the late 19th century). The small fishing port of Inverbervie was the most northerly point on the North British, on its own branch line, without running over the Caledonian.

The railway along the Moray Firth coast would today be a superb Aberdeen-Elgin tourist line if Beeching had not ripped it up (and Scotsmen had not gone and built on it in places since) in favour of the direct Aberdeen-Elgin line (which was GNSR to Keith but Highland Railway from Keith to Elgin). The hilly terrain was very costly to the GNSR, which built a bridge over the River Spey close to its mouth about 6 miles west of Buckie; to this day there is no road bridge there (although the railway bridge can now be crossed on foot). This brought the GNSR to Elgin, already reached by the company from Keith via Dufftown, a diversion from the direct route to serve the famous Glenfiddich Distillery and many others beside the River Spey, and its tributary the Fiddich. West of Elgin the Highland ran to Inverness, its headquarters and origin. Such was the strength of company feeling that in 1921 this mere 100-mile-long line remained divided between the GNSR and HR portions of the LNER and LMS railways until 1948, when Scotland became one British Railways Region, separated from King's Cross and Euston and the rivalry between them, which persisted in England until the Regions were abolished.

Thus the fish from north-east Scotland went by GNSR to Aberdeen, by NBR to the English border, then by the North Eastern and Great Northern railways to York and London King's Cross. Fish from Moray Firth villages west of Lossiemouth (north of and near Elgin, which is on the River Lossie), such as Burghead, Forres (Findhorn) and, importantly, Nairn, went via Inverness and the Highland to join fish trains from north Highland ports and Scottish west coast fish from Oban, and to run to places like Manchester and Birmingham as well as London Euston. The Moray Firth, 40 miles wide at Buckie, narrows rapidly towards Inverness, where it is now bridged also by road.

Unlike in the rest of Europe, where in 1921 international railway standards were laid down for through running of trains, with tight screw couplings and air-brakes or vacuum-brakes, the UK had loose-coupled vans with only a 'parking brake'– operated by a long pivoted bar attached to the brake shoes that could be held in position on the side of the van by a movable pin when the van was 'stabled'. Dropping the bar applied the brake, but this could not be done except when the van was stationary, perhaps at the top of a steep incline; railwaymen had to spend time walking alongside the train to pin down the brakes, and yet more time releasing them all at the bottom of the hill. The weight of all these unbraked vans was really more than could be handled by the brakes on the engine at the front or the guard's van at the rear; the latter was provided with a hand-turned screw brake handle inside the van.

However, the new 'fitted' vans had a continuous vacuum-operated brake controlled from the engine, and were thus coupled next to the engine, allowing the fish trains to be stopped more easily and safely, and also to run much faster than before. They were definitely faster than lorries until 1940, and could carry far more than road vehicles. Thus the railways had regular contracts with the wholesale fish firms of the big city markets like Billingsgate in London, whose agents (often their own employees) attended the ports and – bidding against the fish curers in Scotland – bought the catches from the fishing boats. Most of the fish boxes finished their journey to Billingsgate Market in railway-owned horse-drawn drays from the passenger termini, which had roadways on or beside some of the platforms for taxis and private cars meeting passenger trains. The London Midland Region still had these drays, often blocking London streets, until about 1960, while other Regions had got rid of them sooner in the capital.

To a visiting Sassenach today, the local village harbours around Buckie seem amazingly small, when for the first time one

Old photographs show small village harbours crammed with fishing vessels, as exemplified by this view of Buckie at about the turn of the 20th century. Royal Navy vessels are anchored offshore. *Buckie District Fishing Heritage Museum*

Women carrying their men to the boats at Portessie in the late 19th century. This was done in order to prevent the men sitting all night in boats with wet feet and limbs (they wore thick knitted stockings). *Buckie District Fishing Heritage Museum*

sees old photographs. They show these harbours crammed with fishing vessels – how they were ever squeezed in, tied alongside each other, seems astonishing today to the passing motorist or bus traveller. Much the same could be said for some Cornish villages. In some places there is no harbour – the fishing boats are grounded on the beach. Often this means that there is shallow sea water between the grounded vessel and the shore, which accounts for the women carrying their husbands out to their ships at Portessie, for instance.

Findochty had a marina installed by Grampian Region in the 1990s and this makes the harbour look even smaller. In the past, fishing boats simply moored alongside each other. Even so, to hump a fishing catch across four ships to reach the shore was onerous. No details are available of how it was done, so one must assume it was all done by carrying fish baskets manually. Supplies for the ships reached them in the same manner. The Grampian Region was abolished soon after the creation of the marina, when Moray District decided to become an independent county again, but Gordon and Banff Districts decided to remain with Aberdeen District to become Aberdeenshire. Part of Western Moray falls within the Highlands & Islands Region of Scotland; this has special circumstances due to its terrain, which is almost as large as the whole of Belgium. The former county of Moray ended at the River Spey, so that the part including Buckie is still thought of by the Post Office as part of Banffshire, though this county has amalgamated with Aberdeenshire. Today, to save fuel and time, many Buckie fishermen, who are restricted in the number of days per month that they may go out fishing, land and sell their catches at Peterhead, leave their boats moored there, and motor the 40 miles home to Buckie. Others do the same from Aberdeen and elsewhere, which is why Peterhead is the largest fishing port in Scotland.

The Findochty steam drifter ***Rosedale*** (BCK6), and the Buckie steam drifter ***Betty Bodie*** (BCK400). *Buckie District Fishing Heritage Museum*

The growth of private motoring that has made this possible was in England always well in advance of Scotland. When steam drifters replaced sail, there were only very occasional visiting motor cars seen in this part of Scotland. Many roads were not tarred until the late 1930s (as in Pembrokeshire), because doing so made them a sheet of ice, impassable for horses pulling carts when the weather was frosty, which it frequently is, Peterhead being 250 miles north of the border. The first motor car whose owner lived in Findochty, 3 miles from Buckie, arrived in 1927, some 21 years after the first steam drifter *Bloomfield* (BF 218) had been built there, on a small slip constructed in an inlet called Crooked Hythe. The builders were James Herd and Thomas Mackenzie, two enterprising carpenters who had been employed by William Denny Brothers, of Dumbarton, whose ships were famous on the English Channel for sailing whatever the weather. The best known was the *Canterbury*, built in 1929 for the 'Golden Arrow' London-Paris Pullman service*. Herd & Mackenzie built a further 31 steam drifters at Findochty in the next ten years, 90 feet long and 19 or 20 feet wide. The triple-expansion engines came from Aberdeen, made by John Lewis & Sons, and only the boilers had to come from Glasgow, which is 200 miles away, made by Dalgleish. During the First World War the company built four standard Admiralty steam drifters, and after the war it constructed Buckie shipyard at the east end of Cluny Harbour (entered at all states of the tide from the west end), and Findochty closed down in 1919. Between the wars Herd & Mackenzie only built four more steam drifters, building instead more than 20 diesel-driven fishing boats. In 1927 Buckie had 290 steam drifters registered and owned, the largest number in Scotland.

* More about *Canterbury* and various other Denny ships can be found in *Pullman and the Orient Expresses* by George Behrend, a 420-page history starting in 1872 and first published by the author at Findochty in 2001, and brought up to date. Published digitally.

According to historical surveys of northern Scottish fishing in the second half of the 19th century, carried out under the auspices of Aberdeen University, and which did not cover western Scotland, Buckie led the way in the Moray Firth, only to have its herring disappear, while the north-east ports of Peterhead and Fraserburgh have grown. None of the harbours were big enough for the fleets of fishing boats, and though Buckie was blessed with the new harbour at Cluny, largely paid for by a wealthy local landowner, its completion more or less coincided with the departure of the herring from the Moray Firth.

Now that the railway to Buckie has gone, because it was neither the sole nor the most direct line between Aberdeen and Inverness, much of the barley brought in to augment locally grown barley for the malt whisky distilleries arrives by sea at Buckie, and goes on by road. This helps the harbour survive its loss of dues from fishing boats. The railway in Buckie ran in a deep cutting and effectively divided the town. To the north the main town, with its very many churches, was cut off from the sea and harbour, which were on a level with the bottom of the cutting, in which the passenger station was also situated. The fish went by horse or hand-cart from the market to the station, but today everything has been swept away, the stone sets in the harbourside road that enabled horses to get a good grip in icy weather replaced by tarmac.

Moray Seafoods, which for some years has held the contract to supply British Airways, among others, is located between the road and the edge of the harbour. Prawns, crabs, etc, are still caught locally, but when local catches are insufficient other shellfish comes from a distance, not just from Peterhead and along the Moray Firth coast – Fraserburgh, Macduff, Banff, Portsoy, Cullen, Findochty and Portessie – but also from ports in western Scotland such as Fort William. However, the enormous fuel prices, the highest in Europe, have put the brakes on this. But there is an insatiable demand for shellfish in London, France, Spain

Fishwives at Buckie station waiting for a train to take them to Elgin to sell or barter their fish. *Buckie District Fishing Heritage Museum*

and Andorra. Andorran lorries come to Buckie for it. Demand grew suddenly in France (and the Channel Islands) when the Brittany Coast suffered a terrible oil slick contamination some years ago caused by an American tanker.

Seafood is one of the main attractions of the Brittany Coast, an area that is also the main supplier of Paris, which closes down (unlike other capitals) throughout August when there is a great exodus to various parts of rural France as the inhabitants return to the often poorer country areas where they originated, although they are now less poor with the influx of retired British, Germans and others. Thus Cornish fishermen have benefited from the EU, unlike Channel Islands fishermen – the Channel Islands are not in the EU and seafood for the tourist hotels is flown in from Scotland and also of course Norway (not in the EU either). All these international complications have descended on the British fishing industry; where before the war foreigners were the enemy, taking their share of the North Sea and Icelandic fishing grounds, but not being permitted to land in the UK. Post-war there were the 'cod wars' of the 1960s and 1970s between the UK and Iceland.

Moving some 70 miles south of Aberdeen we come to Fife, a former Kingdom and a very independent place. To the west there is a ridge of mountains, on the north the tidal Firth of Tay, with the River Tay running out to sea between Fife and Dundee on the north bank. To the east, Fife faces the North Sea, and the south faces Edinburgh, separated by the large Firth of Forth.

Fife fishermen had their own style of fishing boats, called 'Fifies', which had square sterns. Moray boats had pointed sterns, and were known as 'Zulus'. Both were larger than the Cornish-built or Shetland-built boats (maybe because there is almost no timber in Scotland north of Helmsdale, unlike in Fife and Moray). The Moray and Fife boats followed the shoals to Shetland and also to Orkney –

Above 'Fifie' sail boats leaving for the fishing grounds. *Buckie District Fishing Heritage Museum*

Below 'Zulus' in Cullen harbour. The story goes that a young Lossiemouth fisherman married a Buckie girl, and because he had always sailed in 'Fife' boats he wanted to follow that model, but his wife's father owned a 'skaffie', and she obstinately held out for that design, especially as she was providing half the money. Eventually a compromise was reached: the husband got his way regarding the bow and his wife the stern, and that, so they say, is the origin of the 'Zulu' – the first of which was built about the time of the Zulu Wars of the late 1870s. *Buckie District Fishing Heritage Museum*

the latter, unlike Shetland, had no fishing fleet – and in addition came south to Yarmouth and Lowestoft. In spite of the lack of electronic aids to navigation, just lighthouses, and no foghorns until the arrival of steam drifters, these north-east Scottish fishermen had been doing this since about 1865, passing the know-how from father to son. It was a dangerous occupation, and ships would frequently sink with all hands, leaving widows and children.

It is perhaps not generally understood that the fishermen's wives worked at mending nets, etc, baiting line-fishing lines, and sometimes helping with gutting the fish all the year round. There was also a large number of coopers, who made barrels from Scots pine (which grows naturally round the Moray Firth, though nowadays the forests are planted and looked after by the Forestry Commission) in which the fish curers stored their salted fish. Having learned their trade, the coopers were also available for the repair of the Spanish barrels that arrived in Britain full of sherry, and which are used around Moray (and throughout the whisky industry) for the maturing of malt whisky distilled in the valley of the River Spey, which runs out to sea 6 miles west of Buckie at Spey Bay (with an ice house at the river mouth). Of course most distilleries are a considerable distance inland, as excellent fresh water is required.

The fish curers were the wealthy element in the industry, and bought the catches at auction in the port fish markets (usually one important curer per port in smaller ports), when they were not bought up by the fish merchants' agents (as mentioned above). The curers engaged the womenfolk (occasionally menfolk too) in gangs (or 'crews') of three – two gutters and one packer. The fresh fish was brought in baskets, called 'swills' , from the quayside market to the nearby curing yards, where gutting-troughs filled with fish and foul brine were set up in the open air. The fish were sprinkled with salt to make them less slippery, and the guts were put in tubs to become fish-manure. The herrings were then

Herring-gutters at the 'farlan' (curing yard) gutting and salting herring. *Buckie District Fishing Heritage Museum*

layered in barrels, the layers separated by salt, which usually came from the sea locally, panned at low tide, or sometimes from salt mines. Most of the salted herring was exported to Poland and Russia. In Britain, the herring was cured at smokeries to become kippers. Many smokeries now smoke salmon, but the local ones closed about 1990, and only larger ones remain. Getting wood chips is not so easy as it once was.

Life was always hard in north-east Scotland, and still is. The church was disestablished, so there was no tithe, but there were instead feudal rights exercised by the traditional families, who were the heads of the clans, and people who had bought their properties still had to pay 'feu', as it was called, even though in England feudal ideas were considered a thing of the past by the mid-19th century. Feu did not take the place of landlords charging extortionate rents, as quite a few did, north as well as south of the border. On the contrary, it was additional, the landlord having to pay feu on his own house and any others he owned, although if they were let out those paying the rent were usually responsible for paying it to the feudal estate of the place concerned. Feu was got rid of roughly about 1990 (not 1890!)

To help alleviate hardship the Salvation Army has long had its Mission to Seamen, but in 1922 it started an Evangelistic Crusade in Yarmouth and Lowestoft, aimed at the Scottish visitors as much as at the Norfolk and Suffolk fishermen, and this penetrated to the Moray Firth, where Buckie and Findochty were always deeply religious. Buckie had at least 11 churches – some have been converted into flats – while Findochty has five. With the arrival of the Depression, the fishermen's families were very hard hit; being so dependent on the sea and quite liable to drown in stormy weather, the church in all its branches has always played a part in fishermen's lives.

One church in Findochty, the Church of Scotland, is superbly built on a hill overlooking the Moray Firth. The setting sun in summer reflects on the plain glass panes (they could not afford stained glass) and lights up the whole interior a bright red for about 15-20 minutes. It is so positioned that it is in direct line with the harbour mouth entrance and can be seen further out to sea than the feeble little light on the pierhead! So if you line up on the church while it is reflecting the sunset, or 'gloaming' as they say in western Scotland, and steer correctly, when the sun sets and it is dark you arrive where you can see the pierhead and its light without hitting any of the many unlit rock outcrops.

Scotland as a whole was hit hard by the slump, and Moray as bad as any – there was simply nothing to eat, as the catch price was so low that much of the takings had to go on fuel. There was considerable unemployment, too, among the coopers and suppliers of ships' chandlery as fewer were required, there being no money to pay for as much as before. The fishermen – and who can blame them? – drowned their sorrows in the only pub in Findochty, and brought home no money for their wives with which to feed their children. Who could blame them? The entire clergy in all five churches, headed by the evangelical Salvation Army, fresh from success in establishing a further branch in nearby Cullen, and with encouragement from the Buckie branch, could certainly blame them. Naturally they were teetotallers, since drinking was wicked in their eyes, and still is in those of their successors. They shut the only pub, and shut it remained from 1929 until 1990. The present pub, The Admiral's Inn, looks out on the harbour and is situated at the other end of the village, handy for the caravan site and the marina, which is crammed with motorboats and yachts that seldom put to sea, but whose harbour dues are only £40 per year!

Fishing revived greatly in the 1950s – the Scottish Office ran a grants scheme for rebuilding the fleet, including vessels lost in war as well as in bad weather – as did the consumption of home-caught fish. The foreign-exchanged-strapped government could augment the strict rationing – even bread, unlike during the war. De Gaulle said 'Non' to the British, who were kept out of the Common Market until 1973, so fishermen could catch as much as they liked. The idea that they were scooping up so much fish that they were destroying breeding stocks occurred to no one, as only relatively recently has the environmental lobby discovered that this was happening. Fishing was only limited by the capacity of the market for fresh fish and the poor prices paid by the fish factories that canned or processed fish, for use in animal feeds, for instance. The notion was that the boats had to go to Iceland because there was no longer fish nearer Scotland, as the fish had moved, rather than that North Sea fish had been almost wiped out.

The whole pre-war fresh fish marketing process lived on, if modified. You caught your fish, then you went full steam ahead for your landing port; if you got there before your rivals, you probably got better prices, as the buyers had to secure whatever amount they were ordered to buy for their wholesale firm, after which they did not care how much fish was landed, or the price – they had obtained what was wanted at the right price to fulfil their orders from above. Other buyers from other firms doing the same thing allowed the auctioneers to get a relatively high price on first lots, and the fish was sold strictly on a first landed, first auctioned basis. It was reasonably friendly, because everybody knew everybody else, and it would all happen again tomorrow. Consequently, if you steamed all the way up the Moray Firth to Buckie to get your fish sold, other boats had long since landed theirs at Peterhead, so Buckie fisherman have for years been used to landing their fish elsewhere, at Peterhead or Fraserburgh, whence they return home nowadays by car. In the past they would go by bus or train, a slow process via Aberdeen, about 140 miles representing two sides of a triangle.

There were, however, very rough and ready guesses about Soviet fish. The Russians had their factory ships handy, and it is likely that some urban comrades would have starved under the USSR if there had not been fish from the wicked West to fill them! So for a long time it was not possible to prove that stocks had reached dangerously low levels – the Soviets would not lose face by admitting that they were short of fish.

Meanwhile Fraserburgh sought more work. It is much further from the Aberdeen railhead than Peterhead, so it built a dry dock, officially opened by the Queen and the Duke of Edinburgh as recently as 1992. Soon afterwards Herd & Mackenzie of Buckie was taken over by Forsythe. Because Moray has some of the lowest rates of pay in the UK, Caledonian MacBrayne sends many of its small car ferries, which ply between the outer and principal islands of Orkney and Shetland, to Buckie for their annual refit and overhaul. In May 1993 Fraserburgh gained a new fish factory. Moray Seafoods continues in Buckie, but another fish factory has closed, its place taken by the Fraserburgh plant.

ALEX HARPER

'...the people today wouldn't do it, they just wouldn't do it'

A former fisherman and merchant seaman, Alex Harper of Portknockie was interviewed in 1988 on behalf of the Buckie District Fishing Heritage Museum by two trainees from the Balloch Trust. By kind permission of the Museum, a transcript of the interview is included here, its original phonetic dialect anglicised for ease of reading. Mr Harper was first asked at what age he left school, and what was his first job.

'Fourteen. I was too late when leaving school to get what you call a berth at sea as the boats all started fishing in May and June, when we got our holidays, so my first job was in the bakers. It was some job, I can tell you – three in the morning until seven at night and my pay was 2s 6d and a 4lb loaf. You would hardly believe that, would you? Of course there was no restriction in the bake-house – if you were hungry you could have got a biscuit or anything like that, but that was the wages.

'I stuck that for a year and then the next year I went to sea in a sail boat, so that was another "out of the frying-pan into the fire" sort of job. In the sailing boat you had to cook for eight men; you had also to keep steam in the boiler and when hauling the nets you had to coil the ropes, you had to go into the coiling boat, put on coal and keep the whole job going all the time. It was just – well, they wouldn't do it now, but that was the sail boat days. Some days were terrible. I remember coming into Fraserburgh harbour – it was mainly Fraserburgh we fished – we were just a mile off the pier, and do you know how long it took us to get in? Five hours – flat calm, not a breath of wind, and that was the kind of life they had in sail boats. You depended on wind, solely depended, unless some good-hearted drifter came along and gave you a tow.

'My earliest recollections of going to sea long before that was when I was four years old. It used to be the custom in this part of the world that the fish curer used to come up before the season, and they used to contact skippers of boats and employ so many boats to land all their catch to this one curer. What happened was that the whole family went to Peterhead, and of course everybody went to the boat with bag and baggage – that was beds, all the utensils, pots and pans – everything went to Peterhead. They went and booked a room and you all lived in that one room for the summer season. The whole season – that was the kind of life they lived.

'Sometimes if the boat that your father was in was fortunate, your certain amount would be landed and then you were free and you could fish for the rest of the season independently, instead of the open market. This was good because instead of coming back in the boat, you got a journey in the train.

'That was how the sail boats went, and I was

The Portknockie steam drifter *Flowing Tide* (BF640) at Peterhead. *Buckie District Fishing Heritage Museum*

in two sail boats. The first one was the *Treasure* and the second one was the *Gowan*, which was, I think, the last boat that sailed out of here. She was sold to Stornoway – they continued fishing after that, but that was the time.

'The worst fishing of the lot was when they finished the Yarmouth fishing. We used to come home here and go to what they called loch fishing, in the West Highland lochs. If ever there was slavery, that was it. You used to put down two lots of nets, say 15 to 20 in a row, and go to the other side of the loch with that, and when darkness came down you piled some nets in the same boat and went out and hauled them in with whatever herring there was and took them back to the boat, and you did this all night long, right through the night, then you hauled the other nets during the day – the same performance, night after night, day after day. I don't know how we stuck it, it was really slavery, and when you had full nets in the boat and four men and sometimes up to half a ton of herring and the water was just coming over the boat... I think that was finished before the war broke out. Nobody goes there now, and it's a good job too.

'The drifters, when they came along, that was a different mode of fishing altogether. It was quicker, much quicker – with speed, you went to sea where you liked. In the sail boat you depended on the wind and very often when you were going out there was no wind, and you didn't get any distance off, and of course you couldn't lie all night and cast your nets and chance your luck, which was very often nothing.

'Later on, as they got better on with the job, you knew what you were doing, you knew what you had to do and you carried on that way; the only difference was that you were sometimes longer away from home. One boat I was in just used to fish Stornoway regularly and you'd come home one weekend in the summer out of 16 weeks. When you came home that weekend the nets were all bundled up – you had to take them off, get a lorry, spread them out in the park on Saturday, get them in on Monday. The womenfolk used to come and gather up the nets on Monday, bundle them up, get them all back on the lorry, down to the boat and laid on. That was your weekend – you were practically working a weekend except Sunday, and that was supposed to be your weekend out of the summer.

'That was the drawback on the drifters, but latterly that was all stopped, they didn't bother spreading the nets in the parks or anything like that. Just sometimes after you came home from Yarmouth instead of spreading the nets in the field you put them on the fences, and sometimes you had to wait two weeks, sometimes three weeks before you got them dried to take them in again. Then it was the women's job to mend all the nets. They used to have them in the house and dust flying everywhere. That was my experience of the herring fishing.

'The Merchant Navy really was the best job, a lovely job, if it hadn't been for the war. You knew everything you had to do and you had certain hours. They were tedious hours, as well – not so now, I think – but in that time you had two watches, four hours on and four hours off. In that four hours you had to eat your food and everything, so that meant you only got three hours, so that was your lot.

'But as regards the fishing, it's different altogether. These nets they've got now, one boat will catch as many fish in one day as the whole fleet would catch in a season – that's the present-day fishing. It's not what we called 'crans' – that's four baskets of herring to a cran and seven crans to a ton – nowadays they'll scoop up three or four hundred tons and perhaps fill three or four boats with the one haul, but it's going to kill the trade. The herring's getting scarcer, mackerel getting scarcer, and the white fish is getting scarcer as well, but it's certainly on the down grade meantime – there will come a day when it will dry up. Peterhead is two million pounds down on last year's earnings, and that's a big drop, but there's always the boats that do well and boats that don't do well, and that has always been the line in the fishing. Some of the boats, even in the sail boat days, used to do well, while others used to manage to scrape through and nothing more. It was the same in the drifter days, but with the drifters, as time got on, the expenses began to kill everything, so I don't know how they do for expenses now with this oil, but I think the oil is going to be dearer than what the coal was. The mode of living for people living ashore is very, very much improved. The womenfolk now have nothing to do with nets or anything like that now – it's nice clean houses now besides, when you used to come home and find dust everywhere.'

What else did he remember about being a cook on the sail boat?

'The first year that I was in the sail boat, oh aye that was monotony, that – one of the crew would learn you what to do, then after the first week you were entirely on your own, and you had eight men to feed, and I can say it was some shambles. Some of the young fellers went in and they made a real good job, but other fellows they got disheartened and didn't finish the season perhaps, then if they did finish the season they were glad to see it over. It was really a tough life. There was eight of you sleeping in not half the size of this room – eight of you had to sleep, and you can just imagine having a boiler stuck in that corner, and my bed was at the side of the boiler, and every night I was just saturated. Well, you did really sleep, you were dead tired, and you slept a certain way, but you weren't rested. It was different in the drifters – some of the drifters had what they called the fo'c'sle [forecastle], and four men would sleep forward there, so that was six, and for the others it wasn't so bad then. I don't know what the boats are like now because I haven't been in any of the modern boats, but I don't think I would like to, because they work day and night and that's a bad thing because there's no person can work day and night and go on for a long time. They go to sea on Sunday night and they steer long distances, 27 hours perhaps, start to work, and the next time they go to bed is when the boat's coming in to land. Well, who's going to be able to stand that? I don't see them able to do it – in time it will tell on them, but that's the present day. Of course the boats are more up-to-date, with all the electric gadgets and everything of that kind – there's every comfort in them compared to what we had.'

Presumably he had no depth sounders or anything of that kind, or a radio?

'Oh no, there was nothing of that, and our bed was just what they call a bag of chaff, but now they've got all their own mattresses and everything far different from what we had. I don't think it's a right thing to say, but it's a good job that passed off because the people today wouldn't do it, they just wouldn't do it.'

How did he manage for weather forecasts?

'Oh you never bothered with weather forecasts, you just had a barometer. Some skippers were good, some were not so good. I was with one skipper who could read the weather just like reading his chart, it was unbelievable. I remember one night in Great Yarmouth, it was a Friday and it was a howling gale of wind, not a boat at sea, and we went out and they said, "He's mad, must be mad!", but we went off to the grounds, we lay for about two hours and the wind started to quieten down, we got the nets out, hauled them in and had sixty crans and came in. The only boat that was out in Yarmouth, and we got a perfect day to pull in the nets and everything, and we had just finished hauling when the storm broke again. He could read the weather, that man, but the majority of them just looked at the glass, gave it a tap and if it went back said, "Oh well, that's not good", but if one went to sea the rest were sure to follow – they were like a flock of sheep.

'That was the bad thing about the fishing – if one went, the rest would follow. All those days are over now, though, and Great Yarmouth fishing has petered out. The last time I was in Great Yarmouth was on holiday,

Steam drifters leaving Lowestoft in rough weather. *Buckie District Fishing Heritage Museum*

Landing herring at the fish wharf at Great Yarmouth before the First World War. *Commercial postcard*

and we went down to the beach and the wife said, "What's the boats doing out there?"

'I said, "They're fishing for herring."

'She laughed, and I said, "They are, that's what they call long-shore boats."

'They used to fish inshore and if the herring was too many in the net they just poured out on the beach, but now what they do, they take the boats in and they have a cradle that goes down and takes the boats right up out of the sea. They don't put them in the water at all now. I spoke to one of the men there, and the herring are completely taken off. He had twenty or thirty herring for his catch – that was the entire catch he had for that day, so it is finished off, so you can hardly believe it. When we used to fish at Yarmouth it wasn't thousands of crans, it was hundreds of thousands of crans that we landed every year. Well, you can imagine, one year there was 700 boats in Yarmouth fishing, so you may guess the quantity of fish that was landed.'

There must have been a lot of girls gutting there at the same time?

'Aye there was a lot of girls. I tried to get that woman across the road to come over and tell you about being the youngest that ever went. When she went to the gutting, if any higher up used to come in the yard she used to hide out of sight because she was under age. Oh yes, they had a hard life, the girls. They used to get tenpence – old money – or a shilling for one barrel of herring to divide between three of them, and in that barrel there would be over a thousand herring, so you may know the work they went through to get their pay.

'When they went to Lerwick they went by steamer. Why they went I don't know, because there would be up to 400 women on board. Married women went as well. They had to stay on deck for the journey as there was no room below for them. The woman I mentioned was one of the few that wasn't sick, and she ate all the others' what they call "pieces" – sandwiches – and everything, and when she was going ashore at Lerwick the captain of the boat called her: "Hey, I want to speak to you."

'She wondered what he was going to say, and when she went up, he said, "The next time you come back take a mouth bag (like horses used to have) with you."

'The home life and everything is different now from fishing. They go away on Sunday night, and the women have nothing to do except wash the dirty clothes when the men come home at weekends, but as for the manual labour of mending nets, that's all away, all gone and good riddance.'

If he got stuck in a storm, what procedure did he follow?

'Well, there was a lot of things, some would stick it out, what we call dodge, put head to the wind and dodge, others would make for the nearest port – usually the majority of the boats would always make for the nearest port – but since this radio came on now they get the gale warnings and are better prepared. In the old days there was just the barometer, but the barometer was far better than any forecasts that you can get – if that glass started to go back you knew there was something on the way, but if they give you a forecast now you don't know if it is going to be right or not. I was just laughing at one feller that wrote in the paper a few weeks back, and he was asking the commentator that gave the weather forecast when he was going to give a correct forecast. You get fellers like, that can always get it a dead right.

'Of course, in the fishing industry now it's completely changed. There was one of those boats – I'm not going to say any names because it's not right to say names – you know what he had for his week's earnings? A thousand pounds per man. Could you believe that? If we got one thousand pounds gross for a whole crew we would have thought what a big fishing we had, one thousand pounds for a week.

'At that time some boats that went away to Yarmouth just had enough to pay expenses, especially older boats, because you got a lot of bad weather at Yarmouth and they daredn't go out in the bad weather. Some of the better boats used to take a risk in the bad weather, then of course if you did take a risk, well, there was a shortage of fish and you got a good price. There were a lot of things to consider in the fishing, but not so now. It's a case of the man that's a good brain – he can work these echo sounders and everything now – they can even tell you what kind of fish swim under the boat.

'Oh, if you could have heard all the girls at the gutting yards in Yarmouth singing! This woman who was the spokeswoman for them, she called the girls out on strike at Great Yarmouth one year. There was a man questioned her. He said, "How did you get on?"

'She said, "I was put in the jail." She then added, "I went to the pictures with the bobby."

'So there was some laughs attached, they were a happy lot. Then in the weekends in Yarmouth you thought Yarmouth belonged to Scotland because all the men were ashore and all the girls were on the street and everything, but now it's all quiet – there is nothing in Yarmouth now as far as the fishing's concerned, it's all oil, all those things.'

What did he do for entertainment?

'Just the pictures and that was all there was. That time we were down in Yarmouth I thought I'm going to have a look round where all the gutting girls used to work. I don't know if it is still there – it's a few years since we were down – but what we called the curing yards have still got the curers' names kept on for what you call conservation. There's Woolcock & Son, Wm Slater and Sons, the curers' names are all still on the yard, but they are all factories now.'

How did he go about ordering food and coal?

'Well, you always had fish-salesmen – they would order the coal for you. The cook had to order all his own food and sometimes they had long treks with a big herring basket on the shore to get beef, vegetables, loaves and everything and carry the lot back.'

How was the beef kept fresh?

'You got fresh supplies every day, you see, drifters landed every day – sail boats as well if they could get in. They never took away big loads of anything like that. There were a few boats here and all along the coast that, when they started out in summer – it was stupid –

Above 'Oh, if you could have heard all the girls at the gutting yards in Yarmouth singing!' *Commercial postcard*

Below Memories of Great Yarmouth's herring curing industry – a postcard franked July 1906. *Commercial postcard*

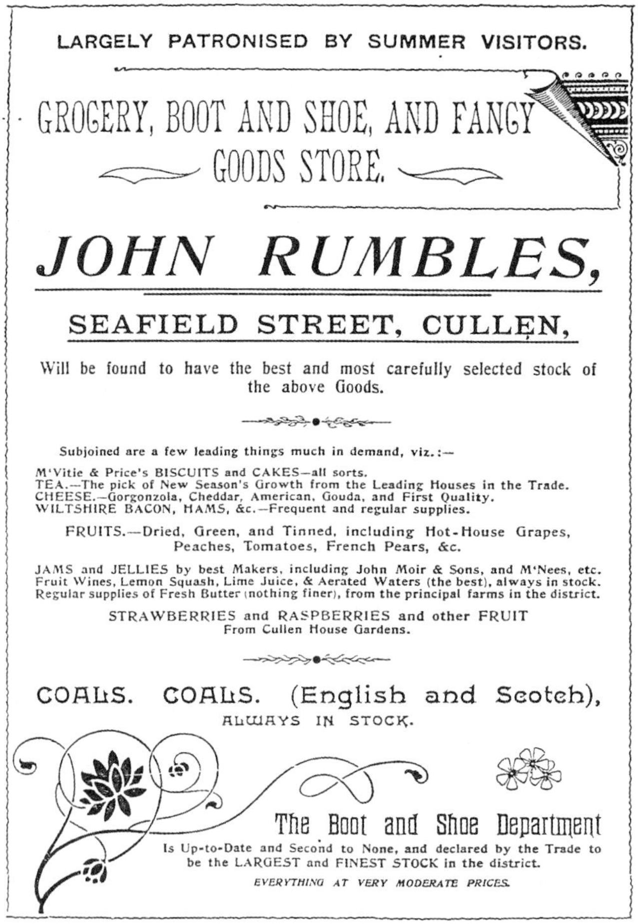

Biscuits, bacon, soft fruits, coal and boots: an advert for a Cullen general store.

they used to give the home grocers a great big order for jam and bread and all sorts of things and packets, and they were only going about 24 hours away – that was old-fashioned.'

How did he manage for washing and drying his clothes?

'Oh, you didn't wash any – you took them home to the womenfolk. You had no time to do that, but in the Merchant Navy you had to wash your own clothes, and it's amazing how you got adapted to those things. I would say that the best years of my life, at sea, was in the Merchant Navy, except for that five and a half years of war – it was terrible.

'There were three boats chartered by the government for stores for the troops, and we used to leave London, go down and anchor in the Downs and wait till dark, then set off for Le Havre. It was the first port of call, then up to Rouen to discharge our cargoes and come back without an escort. We did that for eight months until the collapse of France, and we must have been one of the lucky ones. When we went into Rouen that time they didn't want to discharge our cargo for some reason or other, but the other boats had so much cargo out and they had to stay, then the Jerries broke through. We got out, but another two boats were captured, and some of the crew landed in Spain. I was lucky. I'll always remember that cargo – it was a beautiful cargo of brand new fridges. The Jerries didn't get that lot.'

What was the first aid procedure if somebody was taken ill on board?

'Oh well, you had to wait till you got to the nearest port, but there were very, very few – I cannot remember any cases in the fishing line when men were taken off boats, one or two perhaps, but not many. They seemed to be a healthy lot or hardy. But oh, the Merchant Navy was a different life altogether. I was on what they called the coastal section, Spain and France, Holland and latterly, after the war ended, Germany. I would say that there's nobody in the world to compete with the Germans for workers, far above anybody you ever saw – no work was any bother to them, they were all the same.

'That time when we went there, cigarettes were very scarce – I didn't smoke, but I got my share along with the rest. We were allowed a hundred cigarettes at eight shillings for the hundred. They were Players, so I used to give the German boys a packet of cigarettes and they would have *lifted* the boat for a fag! Aye, a great prize at the end of the war. It was the same with whisky. I could have been drunk every day. Eight shillings a bottle – that was duty free, you see.'

What was the cost of a ton of coal for the drifters?

'Twenty-seven to thirty shillings. It's not real. A hundredweight was just over a shilling.'

How long did it take to get steam up when they were setting out?

'Well, when you started the fishing the boiler was filled up with fresh water. You didn't push it – you gave it about twelve hours just in case it would blow any joint or anything like that. You had to be careful because the boats were tied up between the seasons – sometimes between the winter and the summer you got three months. That was the only good thing about the herring – you had a long spell at home between the winter and the summer.'

Did the bottom of the boat have to be scraped during that time?

'Aye we had to check her right round from stem to stern, three week's working. Do you know what I got for my three weeks? Twenty shillings! That's incredible! It's unbelievable, but it's true. Nobody had the money to give you, and if you didn't get the boat ready then there wouldn't be a job.'

Were the boats hauled up on to the beach?

'Aye, haul them up on the beach, chip all the rust off and paint them all around.'

Did he remember any of the drifters he was in?

'I was in several drifters – one of them was the drifter *Celocia*. I always used to wonder what that name was till one day when we were on holiday I went to Belle Isle Park in Ayr and there was a flower growing it, just like a feather duster, and the name was Celocia; that was the first time I ever got to know what it was. The first drifter I was in was called the *Hearty*, then I was in the *Dexterity*, the *Celocia*, the *Exarantheum*, and I finally finished up in a boat they called the *Fisher Lad*. That was one of the latest, a real class boat that. Of course they're improving all the time. If you went to the herring fishing there was some of them that just went one fishing then away to another boat. They didn't get any better than anybody else – you were just as well to take your chance with the same boat. You had the same opportunity all the time, but that's all by-gone days now – no more nets to mend – and I'll tell you this much. You speak about sneezing when that nets were being mended – everybody was sneezing. It used to be in the summer time especially. I don't know what they call them – we used to call them 'scalders', stinging jellyfish that got in the nets, and when they dried you would sneeze – you couldn't stop. Big things with red tails, those stingers – the blue ones you could pick up in your hand, they didn't sting.

'Some of the fishing you liked and some you didn't like. I never liked Lerwick, because in Lerwick all the curers had a wharf to themselves. When you went to that wharf to land you had to put your baskets on a bogey and push them uphill to what they called a forehand where they emptied the fish in, and before you got that done you were practically on your knees! It was murder! It was fine if two boats got there, then you would have two crews. You would put on four baskets and they would put on four baskets – eight men to push up instead of five.

'There were lots of snags in the fishing industry that could have been ironed out, but I don't think the herring fishing would work now. Nowadays there's too much traffic on the sea. You see in our day you would have great expanses of water, and you would never see a boat, but you wouldn't go far now without seeing ships. There's all that boats that attend the oil-rigs, but there's one thing I don't agree with, and it's this automatic pilot that they've got for steering the boats now. It's a very, very dangerous thing. My nephew has a boat. I went to see her out in Buckie – there was a launch, she was new, and they'd got an armchair and I said, "You're not allow to have that! What are you doing with that?" He said, "Sit down and you'll see." So I sat down just a few seconds, then a "burrr" made you jump up. I said, "You won't fall asleep there." But I said, "That doesn't help – you can't look out sitting in a chair." This automatic pilot steers the same, it never alters, that's why a Buckie boat got sunk a few years ago It had an automatic pilot, and hit another boat. The feller went down to make a cup of tea, and in the time he went down he smashed into another boat. There should always be a man on watch. The sea's different from the land. On a road you can see everything that's coming and everything that's behind and all around you, but on the sea you've got a big expanse of water. You might be looking there and see an object coming this way, and there'll be another object coming that way that you don't see, so you need somebody on watch. It's been very, very lucky that there hasn't been an accident – it will come, because it just can't go on for ever and ever with just one man on watch.'

Was it the case that in the past whole families seemed to go in the one boat?

'Oh aye, there used to be father and sons, uncles and their sons, who used to be the entire crew of some boats. It was a bad thing. I remember the first accident in my lifetime was a boat they called the *Evangeline*. It was during the winter fishing that boat was wrecked, about Stronsay – she went ashore and crashed. The whole crew was drowned, unless I'm mistaken. They were the first burials in this cemetery up here. The whole crew was lost. She was a sailing boat, the *Evangeline*.'

Did some of the sail boats get motors in them eventually?

'Oh aye. Just before the first war finished they were what we called converted into motors, but it was laughable – the biggest engine was a 75-horsepower Garner, the next was a 60-horsepower Kelvin, and you can just imagine one of those big boats being driven by a 60-horsepower Kelvin. They ran on paraffin.'

What was the protection like on the sailboats? Was it easy to be washed overboard?

'Oh aye, there was no rails, not like the drifters, but I think myself now they're going too far with that. The man that lived across the road from us, he was lost in an Aberdeen trawler – I saw that trawler and I wouldn't go in her. From the deck up to the top of the rails was as high as that. Well, you can just imagine a sea breaking in on that – how is it going to get away? I'm sure that with those boats the deck filled with water and when the sea went to that side it capsized. They're too modern with the boats now. Everything is for speed, speed, speed. Aye, speed and efficiency. I would say this – in the old days life was far better than it is now. Your neighbour in the old days was a neighbour, a real neighbour. When we were young, if my mother had a pot of broth on she'd be in with a bowl to next door, and the next day the woman would be in with a bowl to us, that sort of thing, but that's all gone now – you hardly know your neighbours now.

'In my young days we just made our own enjoyment. We used to play football until you couldn't see the ball, all night – that's why I wonder at these fellers saying they play ninety minutes and they're dead beat. We used to play for five to six hours at a time. There was nothing that entertained you in the house unless you had a mouth organ or a melodeon or anything, and if you had that somebody would say, "Be quiet, stop!"

'Sunday was kept as a Sunday – there was nothing done on that day. There was no work of any kind done, apart from just washing your dishes – some people didn't even do that. Now Sundays just like another day in the week. Oh no, there was no movement on a Sunday, and if you went anywhere you walked. There was no traffic of any kind – you'd get the road to yourself on Sunday.

'The churches were packed full. Every church was full – now it's the other way about. I believe that when I was a youngster I would honestly say in this village that if you got five families that didn't go to church that would be a lot out of the village. Its more like the other way around now. They weren't rough, though – they were quiet folk, but they just didn't got to church. That was the funny thing about it. But the churches used to be packed full to the doors every Sunday.'

The harbour must have been packed with boats in those days.

'Well, the most I ever saw in the harbour was when the war broke out in 1914. All the boats had to come home because most of the men were in the Reserves, the Navy Reserves, and were all called up. The harbour was packed full inside and outside as well – that was practically the finish of the sailing boats. There were very few sailing boats after that, but, oh, it was a tremendous sight, all those great sixty-foot masts. It was a beautiful sight to see one of the sail boats under full sail in a fair breeze, a beautiful sight. They couldn't go over like the drifters – they would list down a certain amount then they would go that way. All the time they weren't rolling about they were steady in the water, a far steadier boat than the drifters.'

Did he remember when they built boats in Portknockie?

'Oh aye. I don't remember them all, but I remember when the last one was built down in the harbour there. They called it the *Berry Brae*. The other one was the *Sunshine*, but I think the rest were built for people away from Portknockie. It's hard to imagine that now – it's such a quiet wee village. Although it was different here on Friday night – there was plenty here on Friday night. Smashed windows, with a gang that came in cars. There's plenty of that unfortunately – it's the same in Buckie. I say this, and I mean it: they'll never stop this vandalism and all the murders and things if they don't bring back the birch. The birch rod would finish that overnight. I was with a feller at school – he was one of the real toughs, there was nobody could stand up to him, he could batter anyone down and everything. He was wicked. When he left the school, well it was that that kept him quiet, but when he left the school he was just a real character, just a madman, but he was a fairly proficient footballer. He wasn't a professional footballer, but he could play alongside any professionals today. He used to play for Fraserburgh and the manager used to take him out of jail to play the match and put him back inside again. He was a terrible character, always in trouble, but he just seemed to keep out of real trouble and never got in the mix among the police, but after he left school and that he was never out of jail. That's how he spent his life, in the jail.

'The birch was given in Banff. There were one or two got it here, and in the main they deserved it I would say, but some of them got the birch for nothing or very little. There was a feller in Cullen there, it was a sin really to hit him because he was a softie. You know what I mean? He didn't understand things really and he stole a turnip out of a farmer's field. He got the birch for a turnip. They went too far in that case.

What were prices like in those days?

'You paid threepence for a loaf, and you got a softie – what was called an outside loaf, a long one, with a crust on the outside. A pint was a penny, and if you wanted to walk up to the farms you would get a big pail that would hold a quart or a gallon for a penny. Turnips cost a penny and carrots were nothing – the wife gave you them. Potatoes were cheap, half-a-crown [2s 6d] a bag – you can hardly believe it now. Money went a long way then. Farm workers got what they called a fee – that was sign on for six months, and they got six pounds for the six months. They got their food and that, but it wasn't really food they got – it was just what they really wanted to give them. They had them for six months and they couldn't really leave, but I still say they had a better life than we had. When you think about it, a pound a month just amounts to five shillings a week. You could get nothing now for five shillings, nothing – you wouldn't get a packet of fags for that.'

Were there any superstitions around then?

'Oh, superstitions, it was full of superstitions. We used to live beside where the chemist shop is now and there was an old woman, a tiny woman, and her skirts were always trailing on the ground – there wasn't a skipper that would pass her without giving her a shilling when they were going to sea. Not one skipper would pass her, everyone gave her a shilling – they were afraid if they didn't give her a shilling she'd wish them bad luck.'

What about not sailing on a Friday?

'Aye, they wouldn't do that. I remember, before I left school I used to go to sea in a small boat with an old man. We'd to go down the harbour on a Friday night because this drifter was going away to Yarmouth, and he wouldn't let go her rope before 12 o'clock. We'd to stay down there until 12 o'clock before he would take his ropes off the pier so we would get alongside. Friday had to be finished before he would let go his ropes.'

Was it the same on Sunday?

'We never went on Sunday unless it was an emergency, anything that just had to be done.'

Did they wait till Monday morning or did they go after midnight?

'Oh no, there was no going away at the back of twelve [after midnight]. Oh no, you had your breakfast and everything before you went down to the harbour. No, there was none of that. It was a better life in that way I would say, and there was more friendliness amongst the people. Today there are people who think they're a better class than me, and they won't bother with you, and there's others again that think they're not good enough, and they won't bother with you either, so you're stuck in the middle. But oh, there is a vast difference in the living now besides the old days. We used to get up in the morning and you knew your breakfast before you got up – a plate of porridge and milk, that was your breakfast every morning apart from Sunday. You might get an egg on Sunday, and then you would have to halve it with your brother. That was a special treat.

'People that were well off would have a toilet, but some people didn't have a sink, just a basin put it in the middle of the table to wash the dishes. It was really hard times without a doubt, but it was happy times. Your neighbour was a neighbour. If any of you were ill you could depend on your neighbours – they never let you down.'

Perhaps because you were all the same in those days?

'Aye, everybody was the same, there was very little difference. Perhaps the man that had a boat, if it was a boat that happened to do well, they might be a step above, but that was all. They didn't class themselves as above – they were still neighbours – but it is not so now. Neighbours don't even trust each other now.'

People often say that long ago you could go away and leave your door open, that you didn't need to lock it.

'Oh aye, after we came to live on this street there was a woman in the street there never locked her door, and that was after the 1940s.

She just went to bed and left her door open. You daren't do that now. It's funny how things happen. We never used to lock the door, but we were sitting talking here one night, just like we are now, and our door opened and somebody came in. She looked at me and I looked at her, but as it happened it was our son-in-law from Findochty that was there. I never said anything, I waited till he went out, then I said, "Now the door's going to be locked." She said, "What for?" I said, "That could have been anybody." So she said, "I never thought about that." So when it's dark she locks up.

'But this is a quiet, quiet corner here – we hardly know anything in Portknockie unless somebody comes and tells us. There is never any rowdyism or anything in this part, it's very, very quiet, but if you go down past where the church is and round about there you'll find all the trouble you want. I was speaking to a woman one night and, oh, she was at her wit's end. There was a crowd threatening to break her windows and all the rest. She said, "I've phoned the police three times and they haven't come." I said, "Well you get on the phone to Bucksburn and you'll get all the Buckie police over." That's the only way you'll get them – there's no policeman here, you see, nor in Findochty, but there is a difference without a policeman. Don't tell me there isn't, because the kids go around the street as if they own the place and they can do what they like – no matter what you think and do, they think they can do what they like.'

JAMES ROBERTSON

'It was hard, hard. But still we were as happy as Larry, happy as lords'

James Robertson, also of Findochty, was born in 1910 and was interviewed in 1988 by L. Rickerby and S. Mearns, then trainees with the Balloch Trust. In the main he was asked about his life as a young fisherman at Yarmouth in the 1920s and early '30s; his wife was present and once or twice he would ask her to remind him of something.

'I left school in 1923 when I was 14 – well I was hardly 14, as I sought an exemption to get away to the sea. The first summer I went away with the crew, they had a cook, but I was learning, you see. My father wouldn't let me take it on till he thought that I could; well, you see you had to make food for ten men and coil 12 ropes, thick tarry ropes, every morning. Well, it was worth weighing up. He said, "Oh no, wait till you're a bit older."

'Well, I was 14 just before I started to cook, and I was three years a cook, then I went on to be fireman. I got a step up. Going in the engine room, you see. Then after about three years there I was on the deck. I worked on the deck as a fisherman, clear of the engines, and worked there till about 1935 or '36. We had the two boats: the *Courage* was built in 1933 and I spent five years on her, then I went on the *Christmas Morn* as engineer. This was to Scapa Flow, where we were called up by the Government. We were there a good while, then we got a shift round to Loch Luxsford and soon from there to Greenock. We were to go back after about a year to get a refit and they wouldn't let us go, they said everything was in order. Because of this the mate went up to the office in Scapa Flow, and he created, and was told we would get the sack if we didn't keep there. He says no, we're going for a refit. I was silly – I shouldn't have stayed there. I could have been there the whole war, but you know if you are alone with a crew you do as they do, and the whole lot of us got the sack. I came home and I wasn't long at home before I was called up. I think five weeks, and I was away to Lowestoft in the Navy, six years there.'

Which was the first boat he sailed on?

'I remember her fine, she was the *Conie*, BF937 was her number. She belonged to my father and his brothers. She was a wooden boat. She was a good boat, a good sea ship.

They looked after her well. If she wanted repairs they did them right away. There was no dilly-dallying about it. That was your livelihood. We fished mostly at Wick and Stronsay – those were the two favourites. We fished there a year, then we were round at Scrabster and Stromness. That was opened that year and we had £50 for three weeks gross – aye, that's gross – and my father says, "This is not going to do." We went home.

'You had to take on good gear for Stornoway because it was big herring that you got there; we put on a new fleet of nets and went away there. We'd never been to Stornoway in the summer time, and everybody was dreading it. We were off Tiumpan Head and there was a great big English boat lying there, and he just shot his nets you see, and we came up and spoke to him. We asked if there was anything doing. "Yes," he said, "we had 140 cran today from this water." So we shot away just at the stern, you see, that was on the Thursday, and we got £440 for our two shots on Thursday and Friday. By the time we left there, and that was a big fishing, we had £1,850 for the six weeks we were in Stornoway. Big money then! Oh it was a fortune then. So that just lets you see the difference between areas. We tried Wick round about Scrabster, Stromness, around those waters – the backside of nothing! It was a hard time for my father going to Stornoway because he didn't like it. He says sure as death that's made our day. Of course I was only a cook at that time.'

What about stores – did he have to bring them with him?

'Oh yes, you had to go up every Saturday. I'll tell you what we did. We took away two or three weeks' stores, like jam and all that kind of thing, but not bread and all that. When you went to Yarmouth you got 14 or 15 big loaves and that did you for a fortnight. Then you got hard biscuits. But no, you went up each day for fresh food. You were in nearly every day, and they came down like vultures with runners who were working for the butchers and the bakers to get orders. But mostly you would take a herring basket and go ashore yourself. You always had provisions away from home, like jam, plenty of sugar and box biscuits. Grannies used to bake that bread, but mostly it was fresh. In Yarmouth lads took down your grub for you because it was awkward – it was pretty far from the river. They took it down in vans to you. Oh, they were eager. When you went up the first day you were down to Yarmouth you could have heard them miles away coming down. Runners, you see, young people just wanting the trade. But, oh, you got plenty of feed.'

What were prices like?

'You got a loaf for 11 pence, and it's safe to say that it was 18 inches or more in length. When you went to Jimmy Taylor's he used to bake them, the old father, and you used to take 15, sometimes 20 or maybe more away to Yarmouth, and then you got that box biscuits. And my granny she used to bake the oatcakes, boxes of them. I'll tell you, you couldn't take too much meat for you had no fridges in those days. It was just a wee bit of ice that you put on it. It was fresh meat practically every day. You were going out and in every day from the harbour and it was fresh food every day. You got a fry of herring now and again.'

When he was a fireman, how many tons of coal did he get through?

'Ah well, that boat wasn't too hard on the coal, but I'll tell you she got 12 tons a week, I believe, and that's two of you down the stokehole – one on watch, a four-hour watch. That's steaming in and out every day. Maybe 10 or 12 tons. We used to take in just enough to take us to Shields on the road down to Yarmouth. You went into North Shields, 13 shillings the ton, best English coal – they filled up the bunkers, and I've seen it all round the wheelhouse too. They used to put in what they called the pond boards, and that made just like a big box on top of the bunker lids. Some of them used to go into Shields after the fishing, going home, and take a ton or two each for the house. Thirteen shillings a ton.

What a difference now, eh? Thirteen shillings a ton, that's what it was costing us. It was great.'

What was the procedure if they ran into a gale? Did they receive radio forecasts?

'No, we never got anything. The first time we had a wireless we were caught in a gale. The cook was supposed to take the forecast and tell us before we shot the nets, and he didn't. He forgot, I suppose. He told us after it was too late, but there wasn't a breath of wind. But my father, he was looking into the north, and I can always remember him saying, "I dinna like the look of that sky." This was just before nightfall – I'd just turned in after we'd pulled up the ashes, and got the stokehole clean and the fires – and I went away forward to my bed in the fo'c'sle, I wasn't well in there when she took that sea. My Uncle Dod was there sleeping forward beside us, and he says, "Oh my, I don't doubt she's away to the bottom." It was coming gushing down the hatch – the companionway that you went down to the fo'c'sle. Of course we had a bit of an alleyway before you went into the actual place where you slept, but it was full, it was a torrent coming down. It was nine hours before we could start to haul, and there was nothing but corks and none of the nets left, they were all gone. The corks and the buoys – that was all that we got.'

What happened if anyone was ill?

'Oh well, just go to the doctor. Get the doctor down to the boat in port – you couldn't do anything. You could do nothing, you hadn't wireless or phones, there was nothing to phone ashore in those days. And you had to pay for the doctor, but it wasn't much. Ten bob at the most to see them. But they usually came down to the boat. They were very good. There was a Mission, the Deep-Sea Mission, and there were nurses in there. If a man had to come ashore they would just go into the Mission and they would look after him until he rallied. They were very good at that. The Deep-Sea Mission went round all the different coasts, Yarmouth, Shields, you name it. Up in Scotland as well. They all had a Deep-Sea Mission and they attended to the boats and the men. Then some of them used to have a meeting on Sunday night, and they used to collect in a place and have a sing-song. They were awfully good to you, the Mission folk. Qualified nurses and everything there.

'It was hard in those days. You didn't have the money, that was it. You weren't making the money they used to do. When you were driving a drifter, you got £2 5s a week and you had 5 shillings more for a stamp. The firemen got 35 bob, and 5 shillings for his stamp for his card. That was the wages. But those that had a share of the nets on the boat, they went just on sheer chance. If they got it, they made it. If they made anything they got it, but it was a chance. They hid no guaranteed wage. They couldn't get the "brew" – what we called the dole – because those who had the nets didn't qualify for it. But we got it. The cook would get it, and the engineer and the fireman, you see, because they were stamping their cards – they had a wage. They eventually got £2 a week and the woman had to go from here to Buckie to collect, over to the fishsalesmen's office – £2. It was hard, hard. But still we were as happy as Larry, happy as lords.

'Oh no, they had hard times. When they came home from fishing, up those stairs through to the far end – it's a bedroom now, but it was full of nets. Mending, mending. The women were mending when we were away to the sea. I've seen my father crying to my mother, "Come down out of there, for the love of goodness, biding up there till 12 o'clock at night." Mending nets, when we were ashore. Then after that a lot of the boats didn't go winter fishing – some did and some didn't, but there was never anything, mostly debt. You were always landed in debt. Then the summer time came, 12 weeks rigging out the boat either here or over in Buckie. Finechty [a local name for Findochty] was famous at that time, you know, for the sail for the boats, but eventually it went to Buckie. Well, they wouldn't even pay your fare to Buckie – you had to take a bike or a bus. Twelve weeks rigging out a boat, painting, and never a sausage for it. Not a thing. You had to do that to keep your berth. Not a thing for rigging out the boat. I've seen us 12 weeks going to Buckie every day, home at tea time and home at dinner time if you could get a lift. Some with bikes were all right, but the old boys just stuck it out till tea time with a soft biscuit or a bottle of dazzle. Aye, for 12 weeks we never went out before the summer fishing, rigging out right from top to bottom, painting, cleaning. In where they kept the very coal, it was cleaned and red-leaded – you know, that red oxide for the bunker beside us. The boat was wooden, but all the stocks around were iron. Oh, we had rare fun, I'm telling you. But still we enjoyed it. Saturday night was worth it all when you got tied up at Gorleston or Yarmouth. Of course the old fellers didn't bother so much, but we were always away ashore, and forgot about the rough.'

What did he do in his time off?

'Some went to the pictures. There was a lot of amusements in Yarmouth around the seaside, especially at Gorleston. I remember we went to Scarborough one year; sometimes we went to Shields before Yarmouth, and sometimes we fished three weeks at Scarborough before we went south. Aye, the amusements and everything were going for summer time, you see. It was great. There was two or three piers. One of them was for skating – there was a big skating rink in there – and another one was for the pictures, and dancing on the Wellington Pier. The scenic railway was there. Oh, there was a lot of amusements in Yarmouth. We enjoyed that, being young lads.

'But there was a thousand boats fishing out of that river at the time and there was a strong current flowing and ebbing; you should have seen the men manoeuvre those boats in that river, and not one hardly got touched. They were masterful at using boats in that days. It was between the skipper and the engineer. He got the word to go ahead or astern from the skipper. It was miraculous.

'Oh, there was a lot of amusements in Yarmouth. We enjoyed that, being young lads.' An undated early-20th-century view of Yarmouth beach. *Commercial postcard*

'I remember once that we were right up to the bridge, and my father didn't like that – he couldn't get to the sea soon enough. We couldn't get a berth that Saturday, and the man said we were to tie at what they called the Toll Market. "You can't lie here – go away up the river." We were right up to Tom's house. He didn't like it away up there. We let go at 12 o'clock Sunday night. I was down the engine room taking off my tie, having been ashore, and do you know what time we got to the mouth of the river for boats? Six o'clock in the morning – and it was only a couple of miles. They all let go at once and if the tide was with you – if it was astern – you had to go if the boat was carried by the tide. Goodness knows why there weren't accidents. There were a thousand there – that's counting the English boats and all. The only day they got ashore was when they didn't go to sea on a Sunday – if they were out on Sunday, they got Monday in. They had a whole free sea on the Sunday. If you were out at sea on Sunday that was one thing, but you wouldn't go to the sea on Sunday.'

Were there a lot of superstitions?

'Aye, I remember one we did as a trick. We were fishing in Stornoway. There are lovely grounds in Stornoway that we used to go into on a Sunday for a walk, and we had a little doggie, called James, and of course he caught a little rabbit, a wee thing like this. We were lying beside this boat, the *Heathery Braes* from Portknockie, and we took the rabbit down to the boat. We lifted the ropes where the nets were pulled up at the sides over the weekend, and we shoved this little rabbit in below the corks. They were laying on the nets on the Monday morning and this lad, Arthur they called him, he says, laying down the nets into the hold, "Mercy! It's a rabbit. Look, it's a rabbit! There's no use in going to sea with a rabbit at the nets." You see, the superstitions – they nearly jumped overboard. You see, they wouldn't mention rabbits on a boat for anything. Oh no, no. They wouldn't go to sea on a Friday, or anything like that. They were awfully superstitious. And if they met a minister, that was worse. Some lads carried it awfully far. My father wasn't so bad, but I think they were dubious of a lot of things.

'As I told you, we had been on the *Christmas Morn* and I hadn't been two days at home when I got my call-up papers. I was away down to Lowestoft in September 1940 and I was there till the war ended in 1945/46. I was at the invasion in 1944 and all over the place. I got through it – I never thought I'd get through that lot. We were making smoke for the landing, and went up to where the Yanks were landing right into the beach, five or six boats right in a line, just staggered. We had a great big smoking machine aft where the small boat used to be. Great big eight-cylinder engine, just obliterated anything when they made the smoke. We had three Yanks aboard with us and there wasn't one of them could work the smoke machine – well, there was maybe one, but he couldn't work the thing himself. I says to our lads, "Boys, if he doesn't learn to work that thing himself you'll never make smoke." The reason they couldn't do it was because they were a sick, in a gale of north-east wind. Oh, they were just dead! The whole hold was made up for them with lovely bunks and everything, where the fish room used to be – it was a converted trawler, you see. And we had what you call ten-in-one rations. The Yanks were provided with everything down to toilet paper, great big tins of boiled ham, anything you could mention, they had it. We scored, though – they couldn't eat them, you see. And the trouble was, if they felt any good they were away mixing up this kind of stuff that was bad for them, and they were just as bad again. Oh, we scored off them, but I was glad I got through that lot, that was very, very bad. I don't know how we got out of it – they were right into the beach, both sides the first two days, and then up at that Yanks' landing. I was on the deck – I was mate at that time. I was in the wheelhouse that time.'

Had he ever been a skipper?

'I had my ticket but I never got a chance

with my father and his brothers. My father was still skipper after I went away to the war. I remember coming home on leave before the invasion. I came home and went to see the old lady, that was my mother, and she said that my father had just had 139 crans that day out of Gamrie Bay. I said, are they shooting as far in as that? Aye, she says, there is no limit now that they have to work to. What a fishing they had. Made their fortune. Thirteen, fourteen hundred pounds that week. That was the time of the war. That would have been about 1942, I would say. Mind, they had a drifter. It was the *Lord Zetland*. It was an English boat.

'I was demobbed in September 1945. I came home and I was still on demob leave when my father came up and said, "Are you inclined at all to go to the fishing?" I could have chased him. I was only just home! He says, "We want a cook."

'"Oh," I says, "I'm not going to cook."

'"Ach!" he says, "you can cook fine now. You'll get the same as the deckies and them now."

'I says, "I'm not stony – I've just got my demob money and everything."

'That was just before Christmas. I wasn't intending to do much. Ach, but I went anyway. You know, they had a right good fishing. A big fishing. It was only five or six weeks. But those were different days, the time of the war – they made a fortune at the sea, and they still do it now at the seine net. But they'll kill the fishing with the mode of fishing they've got now. Like these purse seiners. I was watching on the TV last night the Taits of Fraserburgh, great big boats, and if you'd seen them – mackerel, thousands and thousands of them. Nets as big as a football park. Twice as long as a football park. And it's about 600 fathoms deep. Nothing can escape it. But they'll ruin themselves. They'll cut their own throats, I would think so, anyway. And they're going in for bigger boats all the time. Oh, it takes something to run a boat now.'

When he was aboard, what were the washing and drying conditions like for clothes?

'You just had to go down to the engine room, take a bucket and swill out your shirt. Well, we used to do that every watch below with the sweat and the heat. But they didn't even have a place to put their oilskins, and the cook used to go wild. There was a railing inside the galley where they made the grub – they used to hang them on there, and the scales, the herring scales, were over everything. He used to go wild. Some cooks wouldn't let them go down the cabin with their big boots, their sea boots, till they'd washed them thoroughly, because it made a mess, you see. And it was some job keeping a cabin clean with ten men. Herring scales everywhere, but I enjoyed it. I can't understand now how we did it, away ashore on a Saturday as proud as punch, after maybe just a right gale-beaten week. It was very seldom that my father missed a gale in Yarmouth. He was a stormy petrel, I would say, going to sea. He would always frighten you. His brothers were afraid, though James – old James, he never lived that way. I've seen him, especially with the boat if it was decent weather, or half decent, he would go into the wheelhouse on Monday morning and he would take the boat to sea and bring her ashore. It was sometimes impossible to go out and in every day in Yarmouth, but if the weather was half decent in the morning and you were getting maybe not too many big shots that would keep you in, I've seen him a whole week working out and in, with nobody relieving him. I don't know how he could do it. Just that wee while when the nets were shot we got a couple of hours before you started to haul, you see.

'The only lad that got to his bed was the fireman. The fireman, you see, he took the boat to sea, and the engineer took her ashore, so you got to your bed going ashore. The driver, he went to his bed going to sea. They were the only two that got a decent sleep. Oh, it was a hard life, but we enjoyed it. Happy times! You forgot it all when you were ashore on a Saturday night. Down to the drives and the amusements, waxworks, and work all forgotten. Oh yes, we were happy, we were as pleased as punch.

'Skipper gave you three pounds and he said well, boys, what about the present money, who is needing this and who is needing the next. Some would say six. "Oh, you can't get six," he would say. It was their money too. We got three pounds each, and you were laughing at your father. But they could do things that we younger lads couldn't do. My father and me were ashore one time buying presents. We were in Woolies, and my father would go to ask anybody about anything, and this time it was little knickers for the bairns, you see, and Tom, he says rubber knickers. I said what's that you're buying? A bag to take home – knickers, those rubber knickers for the bairns. God, I says, I'd better take some too. I went to the counter and went to speak to the girl, an English girl too, but I couldn't. She looks at me and says would I like the same? I says, aye, the same as Tom. She says, who's Tom? I don't think she knew who Tom was. I says, my mate there. What a laugh the girls had about that. I would never have bought them myself, but my father, he could go and say what he wanted!'

JOHN GRAY

'I didn't care who I spoke to, if I had anything to say, I said it'

John Gray of Buckie, a former chief engineer on a trawler, was also interviewed in 1988, and was first asked for the name and number of his boat.

'Oh God,' he replied, 'there were twenty boats. But I started away out of here on the *Violet*, BF1474. I was cook and my father was driver. It was a steel boat, and the working conditions were hard, I would say. The money wasn't there for a start. I went as a cook for 25 shillings a week the first year, 30 shillings the next year, and I went as a deckie for 10 shillings a week. For two fishings, 22 weeks in the summer, 12 at Yarmouth, I got £9, and I had to buy water boots, oilies, the lot, and the fishermen tried to get a guarantee of 10 shillings a week, but were turned down. You got nothing. The older folk, they had to take their groceries and all from the likes of Jimmy, old Jimmy Paterson – he used to be down on Main Street – and various other grocers had to give you tick [credit] when the fishing was down, and they were down as well.'

How did he order his stores as a cook on a boat?

'Well, when I was cook I used to go up to the grocer in Stornoway. I'll give you a laugh about that. You had to ask the skipper – you see, it was his boat. I says to him this Saturday – we had had a poor week – I says, "What shall I take down for our breakfast?" Says he, "I don't know why you're worrying about something for your breakfast – we've a whole box of salt fish." On Sunday morning I'm chopping sticks for the little stove, then I go forward for two cod, about this size, and they're stiff, like frozen. So I'm hacking away at the fish, but the knife's making no impression. Then Alky comes up and he says, "What in the world are you doing now? Just throw that stuff away." I says, "I'm making the breakfast." "Oh for the love of God," says Alky again, 'you're not putting that out?" He goes away down below, then up comes the skipper. As true as I'm sitting here, the skipper says, "Who told you to put that out?" I says, "You – I asked you yesterday." He was going to blame me, you see, and I had to boil the fish three times. And when you were washing the dishes the skin was all stuck to the plates like glue.

'Before I joined that boat – as I said, my father was driver – things were that bad the crew had peasemeal brose [peasemeal in boiling water seasoned with salt and butter] three times in a day. That's a fact. We used to go away from home when I was cook with about two months supply of beef in a barrel on top of the galley, salted and spiced. It was the same with the groceries we were buying.'

How did they manage for weather forecasting?

'Oh, it was very, very grim then. A wireless and more or less the skipper's own discretion, you know. Oh, there was none of this modern gadgetry. When we left we thought we'd maybe go northwards and if his luck was in up it came, on other days we could go away to the south – then nothing, you know. Oh, it was all practical knowledge.'

What action was taken in severe weather and storms?

'Well, the older the skipper, the more experience he had. They could foresee bad weather. But nowadays they have all the instruments, but you can see for yourself that the forecasts can be wrong. Common sense should prevail.'

What were his working conditions like?

'Awful. I joined a ship in Aberdeen, a trawler, and I've been lying in my bunk reading, and out of the side of my eye – you wouldn't believe it – the bloody rats were up on the table and away with a slice of loaf. At the edge of the table there was this upstand [a raised edge to prevent things from falling off], but there was a space at the corners so that you could wipe down the table. The rats couldn't get the loaf over the upstand, and talk about brains, they took it that way and down the leg of the table. I caught I think it was 18 rats in nine days, and I got the Sanitary down and he got 21 while we were at home. That rats caused more trouble in a boat with their stealing. One old fishermen had a muffler of admiral cloth – well, the rats had it away for a nest, and he blamed everybody for stealing it off him – "Where's my muffler?" He didn't know the rats had it away, and that's a fact. Oh, you wouldn't believe some of the stories I could tell. Aye, that was trawling. But now, since we got into diesels, in my later life, you never see rats. Whether it's the smell of the diesel oil, I don't know, but you don't see rats the same now.'

What about the cold, the wet and the dangers? Were there a lot of dangers?

'Ah well, there always has been, you see. The weather can rear up – where there's not a bit of discretion for seeing something, you can run into danger, you see. Aye, we lost a boat in the Firth, the *Dalewood* I think, and I was that bloody heavy going in over the rescue boat that I burst my ribs. I was about 18 to 19 stone going in what they call a pocket trawler. I was off many weeks, I can't remember now, but a good while. The *Dalewood*, an Aberdeen boat.'

What did he do off duty – did he live it up?

'There was one skipper, he was a one – he's dead and away now, a Buckpool man, he slept below the wheelhouse. One Saturday afternoon we were in a trawler waiting for the football on the wireless, the second fisherman and two or three of the crew, when he came up and he switched it off and he says, "I pay for that, that's for fishing purposes." That's Gospel, as true as I'm sitting here. That's the type of people you had to live with, you know, arrogant, just – it's hard to describe. If you spoke out, you see, that was you sacked, whether you were a good man or not.'

What were conditions like aboard? Did they all get on well or was there any fighting?

'Oh, there was quite good harmony. The food was the biggest problem. When I was cook aboard the drifter I think I did away with about four crans of herring, boiled, fried – that's 16 baskets for a summer fishing. The skipper says, "I'll give you a treat", and came down with a box of kippers. God, I don't know how we survived.

'Once we were down at Iceland in a trawler, an Aberdeen trawler, and the skipper was that bloody afraid that a big haul of fish would break the deck. Those were the conditions down there. A big catch of fish, 60 or 80 baskets, they'd let it go when it was on the deck, but no, I'd let it right down before I would open the bag. That was an old steam trawler.'

What were the cooking and galley conditions like?

'Oh well, that stove would be it. I used to lose my rag with it. My father, having been there two or three years, would say just have patience. When the stove was going it was always smoking, and your eyes were sore – you lifted off the funnel and put down a kettle of water and swilled her out into a basin, and it was away like a daisy. There was a hole in the back of the stove, and it was a new stove at that.'

How did he get his clothes dried if he got soaked?

'Well in a steam job you weren't so bad – you see, it was up to the chief of the boat. He could say "Aye, aye, Admiral" when he was asked, then you could dry your clothes below, otherwise you could leave them wet and put on a new shift. Well, you always carried one or two spare sets, you know. At the trawling you used to take your clothes home at the end of a trip.'

Was there any safety or first aid on the boat?

'Well, all the skippers and mates were supposed to have first aid training. I remember one of our deckies, he had no hair, and the boat took a lurch and his head went into the winch, right on it, and the skipper couldn't look at him, he was sick. He put him down and called me up out of the engine room, and I says, "What can I do, you're the skipper." He was that sick he couldn't face it. Oh, that man, I can't remember how many stitches he had. Now the next skipper he'd sit down and sew this coal bag. That's a fact, that was part of their ticket, you see. They had to have a first aid ticket. The skipper, he says, "I can sew," and I says, "You bugger, you're not going to sew me!" What a useless brute, but he was that full of his own comfort.'

Were any men injured on the boats he was on?

'Aye, I recall there was a man died. He was well-built, not quite as big as me, but he was big, I would say about 18 stone. We'd had our supper and they were shooting the gear away – of course we had to go on deck. I was off watch, and George, the skipper, he came flying out of the wheelhouse and he says come and give me a hand – there's something wrong with the second fisherman. Well, he was a big bloke and all, and it took us all our time to carry him into the galley. He spoke to the skipper twice. I got him turned over on his side and he was kind of frothing at the mouth. We got him on to the galley floor – he was that heavy we couldn't risk putting him on his bed. I was left with him while the skipper contacted the Coastguards to get guided into Westray, then kept coming down to see what was going on once he got the boat under way. "How are you getting on?" I says, "Twelve bloody men on this boat and I'm the only one left with the man," and I'm rubbing his hands, and you knew that he was away. I shouted down, "You buggers, the man wouldn't have harmed you when he was living, and he won't now he's dead." Aye, but they weren't long going into his bloody berth for his fags. Well, that night we got into Westray, and got down the doctor, and he asks what we did. "Ah well," he says, "you couldn't have done anything more – that's a massive heart attack." And there's two or three baskets of fish lying on the deck and the skipper says to the crew, "Go away and square up the deck and take three fish and sell them yourselves, and that will give you some pay."

'You did get irritated with the type of people you were sailing with. Christ, they were going up to the jail at Craiginches for deckies to go away with us. The runner came down one day and says, "I've got a good deckie for you." Aye, this was in Aberdeen, and I swear to God you'd have needed a camera – he had on one of those black airmen's helmets, a black coat like the Germans had, a leather coat, thigh boots, leather boots like they had away back at the Battle of Waterloo, and that was his outfit. Well, they gave that man gear, oilskins, sou'wester, water boots. This was supposed to be a good deckie, but he didn't know what the hell to do. The skipper says, "That man is no use – will you take him down?" Well, me and the seconds had to do his work. We used to have a 5-gallon drum for ashes, you know, but he couldn't lift it off the stokehole. He had to do it half-full, and in end

I says to him, "You come down here and fill the buckets." Talk about an experience – aye, and he got all his pay, too. The skipper and mate never as much as said thank you.

'My brother-in-law was with us once at Stornoway, and we were bleeding the boiler and putting in two sockets. At about one o'clock in the morning, I says to him to go and get a hose and we'll be ready to fill the boiler. He comes back to me and says, "There's a light in the fish room." "Light in the fish room?" I says. "How could there be a light?" There was no power, we had no steam, we had no electricity. I says, "Is there anybody on the deck, the skipper or the mate?" He says, "No." "Well," I says, "go away and tell the skipper." Well, when he went for'ard there were three of them in there – a Buckie bloke and two from Aberdeen, the cook and another deckie. He took off the hatch to let them out, and there were I don't how many baskets of fish and bags of rockwood [a wood-like asbestos]. They were going to flog it themselves. I says, "Go away and get the police." "Oh no, we'll be kept here," says the skipper, "and there'll be an enquiry." When we were ready to leave those three stayed on the quay – they weren't coming aboard. We were slowly getting the steam up, so when we were nearly ready I says to the skipper, "What about turning the boat round so that her head is away out of the loch?" We did this, and after a while we left Stornoway, leaving the three rogues behind.

'Normally there was ten of a crew, so that made only seven now. The skipper says, "We'll try a haul and if we cannot manage we'll go away home." I says, "That's no bloody use. You've a cripple man on the deck there, and a young lad about 17 who couldn't lift a basket of fish." I says, "If you put him in the galley, Coullie and me will do the cooking and the baking and he can dish it out, wash the dishes." We had 21 hauls after we left Stornoway, and that thing in the galley never came on the deck once. So we're nearing Aberdeen now, and those three are home before us. We're getting our pay, and the skipper says to me, "What do you think, then – you're awfully dour about this." I says, "No bloody wonder." I says, "That man has to have a second's ticket to be here, and that crowd that jumped her is away with more bloody money than him." If the skipper and mate had anything about them they would've taken a pound from each of them. He took a pound off the cook. When those three went ashore, the rest were left aboard, and the deckhands got three men's shorthand money plus three men's roes [fish roes and liver were put aside for the crew]. Well, the seconds got none of that."

'Oh my, I've seen and heard a lot, and it was just a sheer existence we were going through. It was only the last five years when I was trawling that I would say I got anything like what I was due. My last trip was nine days – thirty-one and a half thousand boxes – and I've landed in Grimsby sixteen hundred and fifty Aberdeen boxes. Twenty-three days down to the North Cape of Iceland for twenty-five hundred pounds That's a fact. No, it wasn't all sunshine in our day, you know. You hear of big trips now, but there's some poor ones as well. There's a boat there that hasn't had a pay since Christmas – the boat was taken from the skipper last week. Now how can folk exist on that, eh? For the last four days they had no lights or anything, and I says to one of the crew, "Who was your engineer?" "We haven't got an engineer," he says. Now that shouldn't be at sea – that's risking men's lives. They've taken the boat from him now. He's lucky, I would say, in a sense that he didn't lose her.'

When he went to Yarmouth and other long-term fishing, how did he get on? Were they good trips or bad trips?

'Well, that depended a lot, you see. I remember the boat I was on, first as cook then as a deckie – trawl men with old gear. That's no use, it's no use at all. If you have no good gear you'll get nowhere, you know. We went out at the end of the war in the *June Rose* out of Peterhead, and the skipper was a Salvation

Army man. I praise the Salvation Army, but there's a middle way. Sunday morning we had our breakfast and he's on to the wee tract sheet. He used to call the crew in the morning with a bugle. He came down into the engine room, and I was chief of the boat, the driver, and I'm away to do a job on the engine. He says to me, "What are you doing? Do you know what day this is?" "But," I says, "this is the only chance I can get." "You leave it alone." Well, I was that afraid of him that I came out of the engine room and let him have his meeting, but as soon as he was ashore I knew what I had to do. I couldn't have left the engine the way it was – there could've been danger if we had broken adrift. I went down and did what I hid to do. I put the tools away and I was sitting aft – what they call the quarter deck – in the afternoon when the skipper came back aboard. I didn't notice him go down to the engine room. He came up and he says to me, "I've been down in the engine room and you've been down there and worked on it after me telling you no." I says, "Aye, I couldn't leave the engine like that." "You disobeyed my word." I says, "I would have disobeyed you supposing you were the King." I says, "I couldn't leave the boat like that – the insurance would have come down on us. But," I says, "in future I'll do nothing on a Sunday and you'll have to wait till Monday morning and breakfast time for steam because I won't let the fire go out."

'They always called us the "Black Squad". They were away to the kirk, Brethren meetings, he was away to the Salvation Army, but he wanted steam for 12 o'clock at night. Now, you can't take it out of a packet. It's coal – steam is made with coal. You see, that's how much they thought of us. "Oh no," I says, "she'll not be ready for you tonight at 12 o'clock." It was an awful attitude, you know. Well, at the time I'm speaking about I've seen as many as a thousand boats in Yarmouth, and they are at sea first thing, but the steam had to be supplied by the two nincompoops, the Devil's men.'

Did he have much time at home?

'Well, we were away at Yarmouth 12 to 14 weeks, you see. It wasn't so bad in the summer time unless you were away round to Stornoway, maybe eight weeks away, then home for a weekend. But it wasn't a weekend. When you were in Stornoway or Lerwick you hadn't much herring, and you were finished maybe at 5 o'clock in the morning. But when you came to Buckie, supposing you had only 20 cran, you had to land them, you had to bundle all the nets, take them away to the bank, and the driver's blowing down the engine boiler. Then him and me had to go down on Sunday morning to fill the boiler. That was another thing – them away to the bloody kirk with great bloody Bibles, and we had all that to do just for the bare essentials. We had to spread the nets – we used to take them and spread them on the parks. On Monday morning you had to go and gather them, take them down to the boat and lay them out. It was just sheer graft. And we used to work by the boat three weeks a month, for not a penny, painting. Aye, it was poverty all right then. Mind, they played on it and all. One man I was with, he put four of his family through college – he's a boy who's a doctor, some teachers, a nurse.

'At the end of the fishing the cooks took the pound jars that had held jam, washed them and put them into a bag to stow them away. We got tin boxes from Fowler's, like galvanised boxes for holding proper sea biscuits, and when the box was empty we used to clean it out and grease all the outside so it wouldn't go rusty. We came into Buckie one summer, the second year that I was a cook, and I had jars lying on the table and was about to carry the biscuit tins home. "Oh," says the skipper, "leave those. I'll take those." He had about six tins, at 2 shillings a tin. We used to get a ha'penny for the pound jars, and he was away with those as well. That's how bloody greedy he was. I says, "Well, well." So when we're in Yarmouth, you see, and I'm washing up all the empty jars, I throw them in the bucket, aye the greasy bucket. My father says, "Aren't you going to wash the jars?" "No, no," I says, "he can wash the jars – I'm not going to." Was I going to wash bloody jars for him?

'I left my pay that year in Yarmouth. We were ashore down in Scarborough, and three weeks in the dry docks in Grimbsy, and consequently we lost the best of the fishing. We were away home after 12 weeks. As I said, the second-year cook got 30 shillings. The skipper says, "Seeing as how we're in debt, I've kept half-a-crown a week off your pay." That's a bloody fact. You see, if I'd picked up my pay he couldn't have done it. Half-a-crown – that came to the large sum of £1 10s. I says to my father, "I won't leave a brown copper after this." There were two boats there from Portknockie for a whole fishing, and they got nothing – their cards weren't even stamped. How they got away with that I just don't know. They should have been stamping the cards every week, that's the law. But they got away – they made out they were bankrupt, you see.

'But there were good boats, and those that were good, they had to work. You wouldn't believe it, all hours of the night washing down, scrubbing down the outside. The boats were working vessels, but they had them looking like yachts. I recall that year I was on the *June Rose*, one Sunday morning I went down and there was John Paterson – you never saw such a hallirackit [hard-working but reckless] bugger in all your life. A white shirt, it was always white, and a tie, and Mattie Guthrie was as odd. They were laughable. I went aboard the boat and she was the pride of Buckie when she was out of the port. Some folk in Main Street and Gordon Street had her. On all their boats the steel plates were shining. Paterson and Mattie put down what you call "dip boards", wooden frames, something like pallets, and on some of the smoke boats the doors were grained and all. Well, the two of them fell out. He was like a countryman, John, great big hands. He went more on cargo boats. He would never have got a living in Buckie – he knew his job right enough, but it wouldn't have suited him. There were more tins of Brasso went aboard than grub sometimes, you know.

'One thing about the herring fishing, if it had survived, I would say it was a great life, far better than Aberdeen trawling, because you were away to Stornoway, the Shetlands, the Isle of Man and Ireland – it was a big variation. I remember when we first went trawling, before the war, you didn't know the day or the date. On Sunday morning you got ham and eggs – that was the only way you knew. The rest of the week it was fish, and you paid for your own meat. In Aberdeen the cashier paid the bills, and we would pay for our own grub, but you see they were taking it off in the office and they were getting like a 10 per cent rebate off all our bloody money. I says to the skipper, "Did the cook let you see the bills?" He said, "Why?" I says, "The grocer his put down here soup, aye like Knorr soup, one and threepence a packet. Brucie has put down one shilling and ninepence." That was sixpence difference. I'm not kidding, I was right outspoken – I didn't care who I spoke to, if I had anything to say, I said it. He says, "What do you mean? I have nothing to do with it." I says, "You have everything to do with it. It's you that gets the hens and turkeys at the New Year – what the hell do we get out of it?" He says, "If you wait for me to approach Brucie you'll wait a long time for we depend on him for a job." "Well," I says, "if you won't do it I will – I'll go to the bloody union and I'll get the bloody fleet out." I approached the union. I says, "It's daylight robbery – we're not making bugger all between the office and the grocer and the butcher. They're stealing us blind."

'He says, "We're putting in for a pay rise of three shillings a day." "Well," I says, "Why not put in for three shillings but put it towards us and not our grub, and let the owners pay the difference." It ran out about four and sixpence a day at that time. The owners thought they were on to a good thing, you see – our three

Buckie harbour in the early years of the 20th century. *Buckie District Fishing Heritage Museum*

shillings stood year in, year out, but the cost of living was going up and they had to pay the extra. The skipper and mate were shareholders – I says, you can do the same as us.

'A Portknockie bloke was cook with one of the big Buckie skippers – just an excellent cook. And the skipper said to him once, "Why don't you put down a bit of fish?" Aye, at breakfast time this was, to save the expense, you see. Next morning I went to waken the skipper – he was like God Almighty, you know. I've fallen out with him umpteen times. Well, when the skipper was called aft the cook put down two fish for him, and all the rest had ham and eggs or sausages and eggs. He didn't know what to say in front of the crew, but he waited until after, when he was up in the wheelhouse, and he send down the deckie for the cook. "What the hell is the idea of putting down fish for me and ham and eggs to all the rest of the crew?" "Well," he says, "it's you that asked for fish." He says the rest of them don't like fish, but this was all to ease down the expenses.

'There was a Buckpool bloke who was cook on the same boat, and the crew were paying for their grub. Now the owner says, "You're going away down to Iceland – see that there is plenty of food aboard." The skipper is talking about their pay – he says to the cook, you're lucky we don't charge you for the fish that you take out of the pond. Now that's Gospel. Beverley Harrow, that was the man – his father was the owner of the boats before that. Greed!

'Aye, and they commanded your life ashore. Not so much me, but I was with a skipper who had an awful reputation – he didn't hesitate to lift his hand. But the man knew his job, and he was mate on this boat, one of Irvin's. Him and this Beverley Harrow used to get their lunch daily in the Imperial Hotel. Well, just adjoining the same building a little further up in Exchange Street a Buckie man had a pub, Flannies, and this mate, he wasn't married, he was in there carrying on, and by Christ they got to know of it. They beat him when they saw him carrying on with – well, one of those prostitutes, but not doing any harm. When he got home for his dinner his sister says to him, "There was somebody up from the office wanting you." "Aye, well," he says, "if they come up again say I've never been here," and he went to bed. He got down next day to be met with, "You were in the Exchange Bar yesterday? Well, we are not going to tolerate that." He says, "What do you mean?" "We don't like ... er, how do you put it...?" He says, "What I do at sea is your business and what I do ashore is no bloody business of yours." And with that the boy walked out. He had to walk about a long time, though. That's the power they had. That was long before Hitler. It's true what I'm telling you.

'Changed days! I've seen 40, 50 boxes of fish taken off the market, and if you said anything you got the sack. That was daylight stealing. And if you spoke up about it in the office they wouldn't credit it, and if you got over-heated they said, well, you know what to do. I fell out with them on one boat. Our cook got the sack, but he went aboard another boat, the same firm, with a £50 backhander. The cook off this other boat came aboard us, and they were so swindled they told me. I'm up to the office and I started. The manager – he's dead now, Charles Scott – he came through laughing. "What's ado with you, John?" I says, "What's ado?" He says, "You've fairly put the cat among the pigeons." But my case was that that was £100 between those two men – it could've been £200 for all we knew. Down comes the Super – he's my boss – and he says, "What's to do?" "What's to do?" I says, "They'd bloody steal blood!" He says, "How can they do that?" "Well," I says, "one person will tell you that we make £12,000, but the next one will say £10,000. You'll get four different answers. We maybe make £16,000 for all I know – that's not legal." I kept on and said, "You're a Merchant Service officer – can you put 140 tons of coal into a boat that can only hold 100 tons?" He started to laugh. He says, that's a nonsense. But that's the corruption that went on in Aberdeen. Now that they're moving into diesel I don't know, but that's charging the skipper and the mate for 140 tons of coal and there's only 100. The boat could only hold 100. That was just sheer stealing, know what I mean? Then, as I said, if you spoke loud enough – tally!

'I was on one boat for three years. We went back and forth for a whole week into Aberdeen – Coull was down with a doctor's line. They're playing on it, you see – they're staying in Aberdeen. They're getting their day's pay, but it was costing us fares day to day. I turns round to the owner and says, "If you can't give us our fare, give us our books, as we would be as well on the bloody dole." We eventually got away to sea on the Friday. The crew is all there, the skipper, and the runner going up and down to Craiginches to get them coming out of the jail. Anyway, I says I hope to Christ we get away today. I says, this is no bloody use to me, for we would be better off signing the brew [dole]. Ah well, we were away for 16 days, then came in and the Super came down and got the defects. That boat used to go to Grantham for coal. It was a big boat.

'Not a word was said to me when I went into the office. The owner, Andrew King, a proper gentleman – he was a bank manager at Gourock – said, "There's your settling, John, and of course you'll have to sign on." "Oh, well, well," I says. "Were you going back?" I says, "I never said I was leaving." "Oh, but we've got a man." While we'd been at sea they had scoured around and got a man, a chief, and he was being paid, you see. They couldn't see their way to let him go on leave. I says, OK. That was Friday. I came home and got a job in Lossie on Monday. I got ten telegrams within the next fortnight from that same firm asking if I would go as chief on a boat, and that was after they had given me the sack. I says, "Oh no, I've got a job ashore now, so I'll stick it." But it was always to suit them, you see. You were out there and you didn't know what was going on. They're shipping somebody in your place. Well, my brother-in-law sided with me.

'They got an Aberdeen man, a right red Communist, and him and the skipper fell out, and messages were coming from the skipper to the second, and the second had to pass it on to the chief. They couldn't get another chief at the time, so my brother-in-law went as chief. Now, he had no tickets at all – Christ, that lad could have got the sack. He went to the Board of Trade and spoke to Allie Coull, and they had to get a chief then. The funny thing about it was, when he went up in front of the insurance they asked if he had a ticket, and he said no. You should have a second's ticket, they said. "Well," he says, "I'm willing to sit for the two now." "No, only the second's ticket" – and he had to wait 12 months. Now, George Reid, he was caught out, machinery trouble, and he went to the insurance and the first thing they asked was had he got a ticket? He got his chief's and second's ticket at the same time. Now if that's not corruption, I don't know – Alex Coull had all the experience, but they wouldn't give it him.

'How we survived I don't know. We were coming home when our bairns were young at six o'clock at night and away at six o'clock in the morning. I never saw them. It was pretty hard all over. When we had the five of them we used to give them maybe half-a-crown a week, but as they grew up they left school, and were marked absent. Ann, she was the youngest, she was getting 25 shillings now. I'll give you a right laugh. When I was just over 16, and my mother and father didn't want me to go into Aberdeen to the trawling, I went to a farm that was wanting a casual lad. The farmer says to me, "Can you drive a pair?" "No," I says, "but I'm quite willing." "Can you plough?" "No," I says, "but it's a casual lad you're needing, not a bloody jack of all trades." He says, "What would you be needing in wages?" I knew what was in front of me – 12 hours a day, Saturday and Sunday included. I says, "Ten shillings a week." He says, "God Almighty! The last man that I had here, I had him for a suit of clothes and hobnailed boots."

"Well," I says, "You'd better go back and get that bloke, because I don't need a suit." You could always get a suit from the Fifty Shilling Tailors. You could clothe yourself for a fiver, and that was for six bloody months! "Oh no," I says, "you'd better get that man..."

'I left school and started beaching with Jimmy Forbes – I was 14 years old and my first pay was 5 shillings. I got a pair of bib overalls for 3s 11d, went home with the shilling and kept the pence. Now I had to work three weeks for the bloody hobnail boots. Three hundred and fifty hobnails. They were like diver's boots. And that's not the best of it. We didn't get out one day of gales. I chopped sticks all day. When Saturday came I got 4 shillings. He kept a shilling off. But part of me was glad to get a job.

'Do you remember Bill Campbell, who used to work in the brew [dole office]? Him and me nearly fell out. He's a right Tory. I says the bloody Tories were in power when we left school and I stood in Land Street along with the other bairns and watched a wife's bloody furniture taken out and peened [a court or Sheriff's order pinned to the door] by the bailiffs. That was to pay her taxes. Well, we're coming back hard that way. This bloody Poll Tax [he was speaking in 1988]. She [Margaret Thatcher, the Prime Minister] is starving Scotland of industrial land. Would you believe what we're paying here a week in rent? I've a bloody "lollypop" job [school crossing warden]. Teresa [his wife] is paying £21.50 a week. This is the idea they're spouting about how many people would vote in their own council. They're being forced into it. Mandy is paying £28 a bloody week. Mary's boy is paying the same. The sooner they take it into their head to buy their house, the better. We've paid council rents all our days, with nothing to show for it. My God, remember what we were paying in the prefab? They can't take enough off you.'

Had he any superstitions when he was at sea?

'Aye, a good lot of them. Pork wasn't allowed aboard, or salmon. Oh, you couldn't *say* salmon. Pigs were called grunters. Remember how they wouldn't speak of ministers? That was the best of the lot – away preaching all bloody day on Sunday, then come down expecting the boat to be ready at 12 o'clock to go to sea.'

Didn't he get Sunday off?

'No, no.'

They didn't leave port to fish for herring until 12 o'clock, is that right?

'This was their religion, you see. Who got the boat ready? Right at the end of the war I couldn't get back into Aberdeen – you couldn't get on the Polars. I was at the trawling before that. There you are – I had to come ashore. I was labouring. Then I got this job to go away to Yarmouth, and when I went to the boat I waited until the skipper came down. He says, "Are you the man from Buckie?" I says, "Aye." "Have you been aboard?" I says, "No, I'm waiting to see you and hear the terms." "What do you mean?" says he. "Well, I've been told that there's some of the watchmen getting a share, you see." "They're not getting a share here," were his words. "OK," I says, "what compensation are we getting?" "What do you mean?" I says, "Those who are not getting the share, they're getting 10 shillings in a hundred." For every hundred pounds you made ten bob. A bag of wages, you see. He says, "You know as well as me it's the driver's job to collect any fish that fall out the net." [These were collected by the stoker to be distributed between the engineer and the stoker for their own use.] I says, "What chance will I have at Yarmouth?" I says, "Take it or leave it. I came through on the bus, and I can go home on the bus." I took the job.

'On board that boat the skipper used to go below himself as fireman. The skipper and all the deckies were ex-drivers – they were all getting a share, they were looking for the big thing. They were all young compared with me. So we're coming in with a big shot of herring and jokingly the skipper says to me, "When are you going to set her away?" When you got her under way, you see, you got your tea. It was my job to go down below and give her full speed. "Oh," I says, "When I see the land." He took it seriously. "When you see the land?" I says, "Aye." I says, "There's an awful vibration in that steam pipe." I knew what I was talking about, more than the whole bloody lot of them. Said the skipper, "Oh that's nothing – these standard boats are all the same." "Right-o," I said, and went down and went to town with her, and opened her up, and the bloody pipe is going, and it burst. I put a bag round it and went up to the wheelhouse and said to the skipper, "Have you far to go?" He says, "About 10 miles, why?" I says, "That pipe that you said wouldn't burst is burst." "Oh my God, will you manage to get her in?" I says, "I'll get her in," and I says, "You or none of them will tell me my job." I got her in and went down to the insurance and told them. That was how dogmatic they were.'

ISABEL HARRISON

'Everybody was the same and everybody helped one another'

The fishing industry didn't just involve the men on the boats. The communities and daily lives of those ashore were also greatly influenced by their environment and their husbands' and fathers' jobs. Mrs Harrison was interviewed at her home in September 1988 by Mrs Patsy Murray on behalf of the Buckie District Fishing Heritage Society, and recalled her childhood days in Buckie.

'Aye, I was born in this house and I've lived here all that time. I played round here and had a great childhood, far better than the bairns have now because we were fortunate to be born at the time that we were because we had a secure, safe background with grannies, granddads, uncles and aunties that lived round about. When you wanted a piece [a sandwich or bread and butter] you just went into your Auntie Maggie's or into your granny's, and there was always a bit of loaf and jam – no chocolate. The only time we really got fancy pieces was birthdays, when you had a party and used to get iced buns and sandwiches from Paterson's down the town. Do you remember Paterson's? Do you remember Maggie Jessie, and the treacle machine, when you went down with a jam-jar to get it filled? They had a machine that used to let out the treacle.

'Everybody played round about and down at the beach with lamies [pieces of glass worn smooth by the sea], at "kick the cannie, heest the flaggie" [a game similar to hopscotch] – everybody played together. You could have gone the length and breadth of Buckie and nobody would have hurt you – you wouldn't dare walk by the Square on a Saturday night now. Down here there was Rita Bowie, Albert Bowie and Ernie Bowie. They lived at the bottom of the steps, and Mary and Lisbeth Geddes lived at the other side of the steps. There was Jeannie Clark, Isobel Bowie, Jeannie Bowie, Albert John, Robert Bain, Sybil Bain, and then Nannie and Marion Hay and all that crowd, and Ernie Boyne and them were next to Elsie's. Then you came up the road and there was Weelmie King, Walter Ewen, Weelum George, John Geddes and Ettie and Margaret Foam. After that there was Nancy Geddes and her brother. We used to have a joke: "What was the safest house in Buckie?" The answer was the Coull's, because they had a "Gunn" on either side of them and the "Barracks" across from them. We thought that was great!

'Our Jean had a bike and Sandy had a bike and I got one – it cost a pound, second-hand from Robert Jamieson's when he first started up. We used to go up to Jamieson's with our batteries to get them charged up for the wireless.

'When we played you didn't need a lot of money. You played down the beach, you gathered lamies, a tin for your weights with a stone on it and you could play "shoppies". Bits of glass, good glass with a gold or a silver strip, that was a shilling. A smaller bit of the same

Isabel Harrison and her cousin Alex Stewart in 1937, and Isabel today. *Buckie District Fishing Heritage Museum*

was a sixpence, a plain bit of glass was a penny, and then if you had coloured glass, like chemist's bottles and that, that was all the different kinds of sweeties.

'We went right down as far as the old harbour looking for tins for our shoppies. Do you remember the old harbour, the Buckpool harbour? They took old boats in there to break them up. We went to the old harbour and used to sneak on to the old boats that were there to be broken up. If your mother found out you were murdered.

'I remember when the nets with their corks were spread on the beach stones, and we crawled under them, and when they had the boiler there when they boiled up the tar to do their ropes. Do you remember when you got a ride on the cart? You sat on the cart and you thought you were somebody, like when you went on the Sunday School picnics, although we didn't because we were [Plymouth] Brethren, you see, and we used to go up to Drybridge or places like that. I recall once we went to Tillynaught. I don't remember how we got there, but we used to go to Deskford and Drybridge, not too far for the horse and cart. If you were lucky you got to sit in the tail of the cart with your legs dangling over the back.

'Then we went up to the station at Buckpool to get the train to Elgin for the Buckie holiday. That was our annual treat. We got money to pay for you on the boats – we had sixpence. And a walk around Woolies, and you maybe bought a purse for threepence. I remember seeing the girders when we were going over the Spey Bridge and I thought they were huge, enormous, like the Forth Bridge. In Elgin I used to stop at Mrs Officer's shop on the way up from the station and spend a ha'penny. You always got something to buy from Woolies, once on the boaties and then home.'

'Let me tell you about when it was rolling egg day – the fun we had. It was a great day for us. You spent weeks and weeks preparing, and when the day came you went off with your

Buckpool, Buckie, in about 1930. *Buckie District Fishing Heritage Museum*

pace eggs [dyed hard-boiled eggs] round all your chums with your coloured eggs and gave them an egg and you got one back in return. I remember when we had an enamel basin filled up over the top with eggs, all with different colours. To dye the eggs some used onion skins – they were the brown ones – and broom blossom for the yellow ones. Do you remember Cooper's cart that used to come round the streets on a Saturday night? I can remember the smell of that fruit yet – his apples were all polished.

'I tell you what I remember most of all that, which was objecting strongly to being at the school and having to speak English, because that was really difficult. Why on earth we had to speak English at our school defeats me yet. OK, teach us grammar, OK teach us English poetry, but leave us to our own language. Did you do art? The art classes – that's the thing that gets me. We used to get a strip of wallpaper border and you had to draw it. That was creative art. No wonder we were thick. But we got a good grounding, although you were afraid, you were terrified to go to school. If you went to school and you were late you got the strap. And what about the poor bairns going to school with holes in their shoes and everything, and them coming in and taking their shoes off and putting them beside the fires to dry. Some of the classes had spare shoes for the bairns that had none or couldn't afford them.

'The country bairns came down from the country and they were often soaking wet in the winter time. They had to walk all the way to the school – there were no buses then. But they were a lot healthier. When you weren't well, when we had a sore throat, we had to go over to my granny's – she lived diagonally opposite our house here and granny and granddad were the most important folk in our young lives. Anything new, you went to your granny first – granny saw your new outfit before anybody else, oh aye. If you had anything wrong with you, granny said if it was the doctor or if she could cure it first. Say if you had a sore throat, she had iodine, some kind of iodine, and a long brush, and she held down your tongue with a spoon and she had this brush – we were petrified, but it cured us. Then there was the camphor in little flannel bags – a camphor block sewed into a piece of flannelette sheet and inserted into a crochet chain, and you wore it round your neck in the winter time to see off colds, and if you were lucky you got a spoonful of cod liver oil!'

The family was a real unit then.

'Oh aye, we had Auntie Maggie along at Great Western Road. My father used to call Great Western Road "small island", and if you were along the town you used to go into Auntie Maggie's for a piece – you didn't have to wait until you went home.'

Did she often play with all the Great Western Road children?

'Oh aye, I was one of the Great Western Roaders. We did a concert during the war to raise funds for the Red Cross. You see, we had a big crowd down here too, our Sandy and everyone, who used to go out in their carts. Mind, there was a time for carts, a time for balls, a time for skipping-ropes and knives, hiplicks [a game similar to hopscotch, with lines drawn on the sand] and double skipping and French skipping. You got a big rope from your father. You see, my father was at the fishing – he took over from my granddad on the *Prestige*, and he went sailing in between the fishing until the *Prestige* sprung a leak. I was four, and I remember that day as clearly as anything. We went along to my granny's, but when your father was away your mother was mother and father both. You wouldn't dare say no to your mother. If you were out playing and the woman next door wanted a loaf out of McWilliam's, your mother went out the door and gave a roar, and you used to run to McWilliam's for that loaf whether you'd just been to the town or not. It wasn't, no, I'm tired – no, I haven't got a bike. You went.

'The bairns did all the errands long ago, and there were no supermarkets. Instead there were a lot of wee shoppies and you did all the shopping there. I remember Paul's the chip shop next door to Georgie Sandison's, and Mrs McDougall next door – she sold everything, like a general shoppie. Then there was a shoppie in Land Street and Jean's in St Peter's Terrace. We had three shoppies to pass on the road to school. Then you had Mary Reid's in Great Western Road – there were two shops, Mary Reid's and Mrs Williamson's, but do you remember Mrs Campbell's opposite the school? That was luxury. I remember getting a ha'penny caramel cup from Mrs Campbell's.'

Did she get a lot of ha'pennies?

'No, no, that was a treat. We called a ha'penny a "maik". It was a maik, and you could get a sherbet dab or a caramel cup. Or the big sweeties that you sucked and they changed colour – gob-stoppers.'

Did she remember Buckpool railway station?

'Aye, Jockie Bosun had two caps – official caps – he was the porter. Next door to us lived Meggie Bosun. Once we made a slide, with John and Weelum George, from our place over to Paton's, and when we went in for our tea Meggie was out with the salt putting it on the slide for making a noise outside her door! There was nothing on the roads in our day. You could see the horse and carts coming, or when there was a bus you could see how many times you could cross the road before the bus came. There was the Copie van, and Fowler's van – Fowler the baker's van – and in the morning you had your rollies [buttery rolls like flat croissants] delivered. Some of them had horses and carts but most of the time it was boys that came round with the rolls and you hung a bag on your door knob and yours was either put in the bag or on your fanlight. Nobody touched them and your door was never locked – you never had to lock a door or anything like that.'

Did she remember seeing folk sitting with their knitting on the beach?

'Aye, and I remember when the women

Buckpool station, where Isabel got the train to Elgin for the Buckie holiday, the children's annual treat. The station closed in 1960 and the line along the coast from Aberdeen to Inverness has since been dismantled. *Buckie District Fishing Heritage Museum*

used to take their basses [fish-baskets] down into the tide to wash them. I hadn't seen women washing in the tide until I went to Greece last year and there was a farmer's wife washing her sheets, and I thought that was just like our mothers washing the basses. The blankets used to be spread on the gorse bushes on the beach. And all the good clothes were put on the greenie [area of grass], and they went round with a rooser [watering-can with a rose] and watered them as they dried.

'I think the summers were warmer long ago. Did you ever make a tent out of hessian bags? Hessian bags, two broom poles and pins – great! And we had concerts and everything along at the salmon bothy – it was great! Another thing I remember was the walk to the Burnie Gollachy for a drink of water out of the well – Munro's it was, along there. "All who thirst may drink while it runs" was the inscription on the stone. And then there was the Ladies Wally – we went up there for a mouthful of water. And there was the "Sanny Banks", but you didn't play there in your good clothes. You had a Sunday outfit and you kept it for Sunday, then when you were older it was passed on to somebody else and it was their Sunday outfit, and then you had your school clothes and you took them off when you went home for fear of damaging them. It was always home-knitted jumpers we wore.

'There was no such a thing as trainers, we just ran about in old cottees. Many folk today wouldn't know what cottees are. A pair of plimsolls – green plimsolls. I used to go barefoot anyway. I recall getting a new pair of sandals – they were Clark's double-barred sandals – and it was near wartime and my mother was moaning like mad because they were 7s 6d and that was awfully dear, very, very dear.

'I've put all my old photos into scrapbooks and written all the relevant details about everything and my boys have got them. All the old photos of my granny and everyone – they're all there. I've put down everything – who they are, what relation they are. When we were little we used to go with my granny to Rottenhillock [Deskford]. She used to hire Donaldson's taxi. She went up there for a fortnight to stay at this farm for a break, and my Auntie Maggie, my mother and us used to go with her. In the car there was stools, pouffes, fold-down seats and everything – we were all ram-packed in – and we all went up to get the best out of the taxi. This place used to have boarded-out folk, they were like the servants who did the casual work. I remember that because my granny always took up something for the folk. Then we used to go down to Sandend and my mother and Auntie Maggie had their hats, their grand hats, on and their crêpe-de-chine frocks to go to the seaside, and me in a kilt and socks to my knees.'

What kind of food did she have?

'You had your rollie in the morning. The rollies were delivered, and you came home at dinner time and it was soup, and your mother used to make a steamed pudding to fill you up, and custard if you'd got it, then for your supper at night you had something fried, but you always got safties [flattish white rolls]. You didn't get sweeties, no way, but when father came home from the sailing he used to go over to Wrights and we used to get a quarter of Parisian Cream chocolates made by Duncan and a quarter of Liqueurs – we used to call them "bottlies". My father went away for months at a time, and you got this half-pound of sweeties. You could eat them all at one go or you could ration them out. We also got Yarmouth rock – Yarmouth rock and a Bible. Our Uncle Tom always took home a Bible. My Uncle Pat had the *Monarch* and my father had the *Prestige*.

'My Uncle Jimmy, he lived in Buckie, he had the chemist's shop where Pearson's is now, but he had to get out of it for unpaid bills. In those days it was the time of the Depression. They went away in the early Thirties: Jimmy went to work for Boots in Cardonald in Glasgow, and my Uncle Robert opened up a chemist's shop in Paisley. They had to go out of Buckie for want because there just wasn't the money – everybody was poor. When I started going for piano lessons at Miss Thomson's in Portgordon it was a guinea a quarter, but I wouldn't tell my chums I was at the music. No way could I tell them – I would have been ashamed. There was no keeping up with the Joneses in those days – everybody was the same and everybody helped one another. We went to the Sunday School – we had a beautiful childhood.

'When the Coronation came we had all the flags – my father brought home his flags and they were strung from our house across the road. You could touch the flags, they were that big, but nobody pulled them down. There was no vandalism – everybody was the same.'

Did she help her mother in the house?

'Oh aye, we all had our jobs. You didn't get out to play until you had done what you were supposed to. Mothers worked hard. My mother and auntie had to go down and do my granny's washing. I remember wash day clearly. The boiler was on, and you had left-overs from your dinner on Sunday for your dinner on Monday. Then Tuesday was the day when they washed the good clothes because they were to bleach, and they had to wash them and hang them out. Oh aye, the smoke could be seen coming out of the sheddie. I remember my granny having bars of soap right round the sheddie hardening, and she used to grate it for soap instead of having soap powder. It was Sunlight soap and she used to buy it in blocks – it used to be along the wall and underneath the roof to harden.

'A lot of folk long ago were really clever but because of lack of cash they never got to go and further their education. My Granny Stewart would have been an awfully clever woman; she used to write letters for folk. She would read their letters for them that they maybe got from their sons in America, Canada or New Zealand, and then she wrote back to them. She was born in a house in Main Street across from the Harbour Bar; my granddad was also born in that street. She went to the Public School, then she left and she couldn't go any higher. I would have said she might have been the kind that skipped classes – she would have been awfully bright. She would have been the one that they got their brains from – some of her family were very bright.'

Did her granny have a big influence on her life?

'Oh aye, a very big influence. You see, my granny and my granddad were Plymouth Brethren. My granddad and Jimmy Duncan and Willie Rosie would have been the founder members of the hall – the building of the new hall in West Church Street. My granddad was treasurer and we all went to the Brethren until my granny died; I would have been 16 when she died. She was in her seventies. My granddad had TB when he was young and he was ill for a whole year, and my granny nursed him. She took out the window upstairs – she got the whole window taken out – and his bed was dragged up beside the window so he would get all the fresh air. There was nothing on the floor, just a chair, and my granny slept with him and looked after him and he recovered enough to go back to sea. My granddad would have been over six feet tall and they had a family of eight. My Uncle Jimmy had fallen when he was little and he had to have his leg off when he was just a young man; he had to go to Southampton to get his leg amputated. When he was little and on crutches him and Willie Smith were at a launch – when there was a boatyard opposite here – Smith "Bodie" and Jimmy and Willie were beneath the trestle table and as the whisky bottles were emptied they were put under the table and Willie and Jimmy they drank up the dregs and were found on the "Sanny Banks" paralytically drunk and my granddad had to cart him home, crutches and all. My granny told me that.

'Round here, at the corner, there was Jimmy Dodum. He used to repair shoes and when they needed soling and heeling you just went round to Jimmy Dodum and he did them. You didn't throw shoes away then. They were all heeled, soled and heeled. Do you remember the tackets [hobnails]? They used to make you slip. You could kick them on the ground to make them spark. They were good for slides.

'Do you remember "Forty Pockets"? He was one of the going-about men that we used to have. He always had on two raincoats and two or three jackets. Then there was Big Joanne, with her basket. She used to come in and say, "Have you got an old jacket of your man's, missus?" Then there was "Silk for Blouses". He sold dishes. I liked "Silk for Blouses" – he gave me a coffee set for nothing. I used to run and take his hand – everybody else was afraid, because he was coloured. He would have been the first coloured person that came to Buckie, then Ali came after that, Ali Mohamed, a photographer. He first came to Buckie with a pack, you see, and I wondered if he was any relation of "Silk for Blouses".

'Travelling families used to come to Buckie some time in the year. One family had their tents pitched up near where the pork factory is now, below the "Sanny Banks" at the back of the railway. Another family used to take up residence where the swings are now, in the play park. There was an old bow factory there at the time. The two families didn't always agree, and when they had too much drink there were often fights. We used to like watching their carrying on, but you used to get told off by your mother for going anywhere near the camps, because you weren't supposed to be there. It was out of bounds, but, oh, it was great! And if you went down the beach you used to be told about Meggie Weerdie. Were you ever told about Meggie Weerdie? We were frightened of her. Somebody would say, "Here comes Meggie Weerdie", and you would run like the clappers!

'A friend of mine, Nannie, was just saying what great times we had playing and making shoppies. You see, she had a sister and I had no young brothers or sisters. The great thing was to have a baby to take in a pram. I used to take out the twins. I don't know how they're still living, the poor things!

Were there baths then, or swimming baths?

'Oh, no, no. You just went into the rocks, and most of the time it was an old semmit [undershirt] and knickers. Some didn't have running water till after the war. There were the walls [wells] on the roads – the taps on the road. Maggie Jessie had one because they had no water. The wells were every three houses or something like that. There was one opposite the corner up here beside Daisie's, and there was one down beside where the swings are beside where Maria Slater lived.'

Isabel was complimented on having such a good memory.

'That's what life's all about – so much now is forgotten. My Uncle Ake made our first wireless, then we had a five-valve Phillips wireless. My Uncle Ake was making wirelesses when we were young. He was good with his hands. He was making crystal sets and everything like that, so that's how we had a wireless.'

So entertainment didn't cost much in those days?

'When we had our holidays we used to go down to Peterhead to Auntie Nell's. She had her crowd and there was all of us and that was brilliant, that was really good! You got a new doll's pram for tea coupons. When you see them now, they have a fortune spent on toys and computers and they don't play with them, they're sitting watching the goggle-box or the video. They don't read or play or make anything. Your brothers had time to make carties and you got a ride in the cart, and when they were old enough they went as errand boys – milk boys or paper boys – and you got a ride in the basket of the bike.

'It's all changed, and when you look round you have to ask, are they any better? No, they are not.'

Didn't she think children were happier now than we were?

'No, because if we had cheeked back at anybody older than us the way bairns do now, we would have been taken in and your knickers would have been taken down and your bare backside would have been skelped [slapped]. Not a shout and a threat – you would've got a hiding. If you went to school and you cheeked your master or teacher of any kind you would have got the strap, and if your brother came home and told your mother that you'd got the strap you would have got a hiding from your mother for being rude. But you were never locked away – you got a row and that was it. Aye, your hiding was done and you knew you was going to get it and that was it. All this "Go to your room" and rubbish like that – psychology. A lot of rubbish.'

SANDY SUTHERLAND

'...one of the ropes came out and put two of my crew over the side'

To return finally to the present day, fishermen in the north-west of Scotland are as concerned about the current state of the industry as their counterparts in Cornwall, and elsewhere around the coast of Britain. One of the men I spoke with in Buckie was Sandy Sutherland, who stopped going to sea in the *Internos* ('Between Ourselves') about 15 years ago. He told me that he had enjoyed his life at sea, but that it was hard when he first started – all that they worked for was food and shelter.

I told him that in Newlyn I had been talking to two young men who fished very close to shore, in fact so close that they were able to sleep at home every night. They had also told me that they had only ever worked under the umbrella of the EEC, and were quite happy with their lot. I asked Sandy if there was anyone in Scotland fishing like that today.

'Not as such, but there was when I was a youngster. At that time there was a lot of fish here in the Moray Firth, and off the coast up in the north of Scotland. That fish has gone now; we have some small boats that fish for prawns inshore. Last summer there was very good fishing for squid – we are all keeping our fingers crossed that they will turn up again this summer. That would give the smaller boats a good boost.

Right Sandy Sutherland. *Michael Charman*

Below Sandy Sutherland was skipper of the *Bezaleel*, which went sprat-fishing for three winters with his partner Willie Hay's boat *Lode Star*. There was so much in the net that day that they filled the hold of the *Lode Star*, then the decks of both boats. That fishing no longer exists. *Buckie District Fishing Heritage Museum*

'I used to fish in the winter, but always nearer home. When I was a youngster we would fish for cod here in the Moray Firth, and haddock as well. That has all gone now – you have to go further and further out. That's why fishermen have had to build bigger and better boats because they are out in all weathers.'

Had he experienced any dramatic events during his life at sea?

'I remember when I was 17 we sprang a leak off Peterhead coming ashore in September. We had the lifeboat that time, but fortunately we managed to make the harbour, although we had to pump all the way into shore. There was another time when I was skipper, and we were fishing 5 hours from Buckie, about 40 miles. We were north-east-by-north and my crew were gutting on the starboard side. It was a fairly rough day and one of the ropes came out and put two of my crew over the side, and that was a pretty frightening experience. We were able to cut the ropes and get them back on board, but by this time they were suffering a bit of hypothermia due to the very cold salty water.

'There was another time in 1981 when we were fishing on the *Internos*. We had two good days fishing then our power block broke so we made for Peterhead to land our catch anyway. I told my brother, who was in as well where we got the fish, so he sailed with us. It was a clear night – you could see for miles. The watch was set and I was away to my bunk. About 4 in the morning I was awakened – it was a nightmare. Everything in the cabin came out, and I didn't know what had happened. I rushed up to the wheelhouse and saw a red light going down the side. What had happened was that there was a local lad in the wheelhouse, and this other boat was coming and making for Peterhead – just the two of us there was, no one else in the sea. He hit this boat, and sank it. I radioed to my brother right away and said that I thought we'd sunk this boat. I stopped and couldn't get a light on, but I knew that I was damaged. Then my brother came on to me, and said that he could see a light in the water. She was a bigger boat than we were, and the first life-raft that was launched wasn't secure and she went away. We were opened up forward almost to the waterline. They got another raft into the water; the stern was up and two of the boys walked off the stern into the life-raft and my brother was able to get them aboard his vessel. We were able to come home at half speed.

As I had in Cornwall, I asked Sandy about his religious beliefs.

'I'm not a deeply religious man,' he told me. 'I don't believe there is such a thing as one religion. I was brought up in a very religious family, and there must be something at the end. I know that. If you read books, for example the American Indians, they have their happy hunting grounds. As far as I can see it is up to every individual to worship their maker in their own way.

JOHN R. ROBERTSON

'A fisherman's job is to fish, not to dump it all back into the sea'

John is also a retired fishermen, and lives in Findochty.

'I started in 1943 – left school in March and went to sea in May of the same year. I was 14 years old. It was drift-net fishing, just for herring, which isn't done now. The last time I did herring fishing was in 1958 – it has gradually gone down since that time. After that I went white fishing, about 100 miles from Aberdeen.'

Had he enjoyed his life as a fisherman?

'Well, the way things are now I would have changed it. Up to the time that I stopped I enjoyed it. All in all, it was good enough. The quotas and all the bookwork were just beginning to take affect the last year I was at it. Now my son, he carried on after me, but not in the same boat. They accepted the quotas

Right John R. Robertson. *Michael Charman*

Below Findochty in 2004: 'I can only think of one boat working out of the village today.' *Michael Charman*

and the logbooks just as a way of life. We really took hard to it. We were being boarded by the fisheries boats, but not with the same severity as it is now. What's going on now words couldn't describe it – it's a wonder that the skippers going to sea now don't have breakdowns. When you're in the wheelhouse you have enough to worry about with the poor prices and poor fishing. I think that what is going on now is a scandal. A fisherman's job is to fish, not to dump it all back into the sea.

'Last evening in Findochty I went to the little cove where they used to build boats – how they managed to build any there is beyond my comprehension. There would be two or three being built at the same time. The last boat built there was my father's, the *Courage* – that would have been about 1933/34. In my time there would have been 10 or 12 fishing out of Findochty; they were converted sailing boats with paraffin engines. Those were the days. I can only think of one boat working out of the village today.'

I learned that at the moment there are only

45 boats working between Banff and Nairn, while in John's time there were 200 out of Buckie alone. I expressed amazement at the very young ages that boys went to sea.

'One time, when we were at Stornoway, there were 17 youngsters aged between 14 and 16. We all went to the pictures. We left about 10 May and were away about four weeks.'

JOHN ANDERSON

'My job was to do with the different ways of freezing fish'

'I first went to sea after school in the evenings and during the school holidays with a man to catch lobsters. He had a naval cutter. He discovered a torpedo off Findochty and was rewarded with this boat. It had a sail but no engine. I received sixpence a week, but the money was nothing really – it was more pleasure for me, it was an honour really. That would have been in 1936. When we came in I would go off and try to sell the crabs – all the lobsters went south. I had a barrow made of a Tate & Lyle box and pram wheels, and the boat-owner had a dog – an Airedale terrier – who had harness put on him. We would take the barrow from the harbour right up the hill to the station – he pulled and I pushed. We kept all the lobsters in boxes among the rocks with sea water running over them.

'I came from a fishing family. My father used to go and sail, and at 19 years old he was skipper of a "Zulu", the *Benjy*. They went to Yarmouth. My brother Willie was a Marine Superintendent Engineer – that tells you what kind of family we are.'

Indeed, John's uncle was Willie Mair, former Chairman of the Buckie District Fishing Heritage Museum, who had recently died unexpectedly. He had been a successful fisherman during his years at sea, being a skipper and owner of the MVs *Elegant I* and *Elegant II*, and his experience of handling a ship and a crew had given him an edge when it came to committee matters! He had been a man of sincere religious beliefs, and the esteem in which he was held in the community was evident at his funeral, where his own Church and the Baptist Church next door were full to overflowing with people paying their respects.

John Anderson. *Michael Charman*

Unloading herring from a 'Zulu' to a Dutch 'Klondyker'. *Buckie District Fishing Heritage Museum*

'From 1956 until I retired, David Flett, of Findochty, and I had two boats, the *Planet* (BCK 50) and the *Seaforth* (BCK 177), fishing mostly in the North Sea for white fish. We were schoolmates in the 1920s-'30s. I finished going to sea in 1986. I had one or two accidents. I started on a drifter fishing for herring. When 12 June came all the drifters and all the other boats put to sea and fished off Lerwick, Wick and Fraserburgh – that was what we would call the summer fishing. The herring were ready for curing and for going to the continent, Russia and different parts. The boats would go south to Shields and Scarborough, to Grimsby, then you had the East Anglia fishing. The were 1,100 boats in Great Yarmouth in the autumn season prior to the Second World War.'

I suggested that there had been a lot of trouble in Cornwall years ago. The Cornish wouldn't go fishing on a Sunday and the Scots were coming south, but he contradicted me.

'No, no. The Scots never went fishing on a Sunday – definitely not. This is where the Englishmen had the advantage over us – they went away on a Saturday night and fished on a Sunday!'

During the war John was in the Merchant Navy, and passed through the Straits of Dover at the time of Dunkirk.

'There were a lot more lost in the Merchant Navy than the Royal Navy – 6,000 were lost in the Royal, but 30,000 in the Merchant Navy. Those are official figures. I was in the Merchant Navy until 1947. After that I was a steam engineer for the CDC – Colonial Development Corporation. I took my ticket for that. First I was based in the Gambia, covering from south of the Equator to the north of Morocco. My job was to do with the different ways of freezing fish. There was a factory ship called the *African Queen*. If you freeze lobsters or crayfish, they go rotten when defrosted, so we were working on that problem. Blast-freezing worked and kept them fresh. There would be rows and rows of sharks fins laid out on the deck of the *African Queen*. At that time I think they got £100 per ton. They went to Japan to make shark fin soup. The carcases were thrown overboard.'

John had clearly had a very interesting and adventurous life for a Scottish fisherman – almost enough for a book of his own!

HUGH PATIENCE

'The difficulty as I see it is for the young men – how do they enter the industry?'

Hugh is, at the time of writing, Chairman of the Buckie District Fishing Heritage Museum, having succeeded the late Willie Mair, mentioned above. His career had been spent as a fish salesman, providing another perspective on the complexities of the fishing industry.

'A fish salesman,' he explained, 'sold fish mostly by auction. When I started in 1949, herring was allocated because at that time it was under the Ministry of Food. But in April 1950 everything went by auction. When the herrings came ashore the skipper or mate would select a sample of the catch, bring it up to the saleroom, and there it would be auctioned. The sample would contain about 100 herring, which would represent the quality of the catch.

'All the fish would be laid out in boxes in the market. Haddock, cod and whiting would be first, then all the flat fish after that. An auctioneer would auction each box. In Buckie there were about six firms, mostly part-owners of the boats, and they would take their time in the market to sell the fish to the buyers. A lot of the buyers represented firms in Aberdeen and further south. The fish would always be taken to markets or to be processed in different areas down south.

'From the fish-selling side of things, only when the fish was sold would the docket from the market be taken up to the office and the invoices prepared and given to the buyers to check that the buying for that day was correct. At the end of the week a statement for that week would be issued to the buyer and he would be allowed until the end of the following week to pay for his purchases – he would not be allowed to purchase any more fish on the market until his bill was paid. This didn't happen often, but would then be taken up at a meeting with the fish-selling companies. They would then decide whether to keep the buyer on the black list or give him a little more time to cough up.

'At the end of the week all of the boats'

Right **Hugh Patience.** *Michael Charman*

Below **Early days of fish-selling: laying out cod on the pier at a Moray Firth port.** *Buckie District Fishing Heritage Museum*

Coopers sealing barrels of herring.
Buckie District Fishing Heritage Museum

catches would be put into a statement for the skipper. Sometimes the skipper would just take the money home and settle the boat himself, but in most cases the skipper would settle in the office. The statement would be shown to him, together with all the expenses involved in the trip, including fuel, food, and landing dues, and so much would be taken for insurance and other running expenses for the boat. Then the net amount would be divided in half – half would go to the boat and the other half would be divided between the crew. So each week there would be a settlement and the boat's earnings would be divided accordingly.

'I started in 1949 and went to Stornoway. I was demobbed in November, and my dad, who was a fisherman, said that Macrae and Duggie Macpherson were looking for help. I went to Stornoway for the winter, but it became a job for life. I was just 20, and I went from Stornoway to Ullapool, Gairloch, and Fraserburgh. I finally managed a firm in Buckie in 1982, and was there until I retired.

'I was brought up further north at Avoch, a fishing village north of Inverness. They still have a number of boats there, but at that time, before the war, there was no official pilotage and a lot of the fishermen made a few bob by going down the Firth and catching some of the vessels coming up, piloting them up to Inverness. My grandfather had a small pilotage cutter, and when he died in 1940 he left the cutter to me. It was a sailing skiff, very fast with a long keel, and very stable – it sailed very well. From that I learned how to sail a boat. It resembled a salmon coble, the kind that was used to fish off the beach. They laid out their net and waited for the salmon to hit the net, then pulled it ashore. He left me the boat and that gave me a taste for sailing. I've had sailing dinghies and after that bigger boats, sailing cruisers, right up until last year. I sold the final one last summer. The family decided that I had had enough – they were a bit worried about me being out on my own.

'Certainly I would say up until 15 years ago fishing was a very dangerous operation. You can imagine working in all weathers and very little protection on the deck. Then they began building boats where all the working operations were under cover, so that took a lot of the danger away. But you can't take away all the danger as boats can be overwhelmed by the sea. I can remember very nearly every winter a boat being lost or a man being lost, but thank goodness that has changed somewhat. There is a training scheme in Fraserburgh that does a certain amount of training on shore, then they go aboard with a skipper who trains them again. At each step the boy is assessed, then there comes the day when he gets his certificate. That again has taken a lot of the danger out of the job.'

If he had his time again, would he still go after the same kind of job?

'I was very happy in my job. A lot of my friends were fishermen, my family on both sides were fishermen, so I was really among my relatives at various ports. Because of the herring fishing you travelled around the various ports where it was being pursued, and you were among the same men more or less the whole time. That was the difference between herring fishing and white fishing.

'I think the decline in herring fishing started probably about 1970. That was due mostly to pair trawling. Really the stocks were being wiped out by 1977. The Government had to do something, and they put a blanket stop on all herring fishing at that time.

'Today there is a quota set each year for herring and mackerel. They mostly go north in Norwegian waters and off the Shetlands, and later they go down to the South West, to the Cornish fishing. Ever since fishing began there have been fluctuations in fortunes – if we go back to 1792, herring fishing started in the Wick waters because the cod fishing had failed.

'I know a number of skippers who have said that the paperwork got them down. The difficulty as I see it is for the young men – how do they enter the industry?'

The Buckie Seamen's Memorial

The Buckie Seamen's Memorial is housed in a picturesque little late-19th-century chapel at the East End of New Street. This was built by the Rev Alexander Miller, minister of Buckie Free Church, to accommodate local services and prayer meetings, and is now owned by Buckie South and West Church.

The Memorial is dedicated primarily to the seamen of Buckie and District who have lost their lives at sea, but may also include others from the counties of Banffshire and Moray. Men and crews lost post-war are commemorated on wall-plaques inscribed with their names and the names of the boats on which they served, while those who perished earlier are recorded in a 'Book of Remembrance', which lies on the table.

The official opening was conducted by the late Mrs Jean Miller on Sunday 4 July 1982, and this marked the end of 18 months of planning and work by the committee that had been set up for this purpose.

On the following day, Monday 5 July 1982, Her Majesty the Queen and Prince Philip paid a visit to the Memorial, and stayed for a time to speak to relatives of some of those whose names are recorded inside. Before leaving, the Royal couple signed the Visitors' Book.

Many people in the district gave freely of their time and money to enable the setting up of the Memorial. Its current upkeep depends entirely on financial support from the general public, and donations towards this are always welcome.

Special recognition must be given to two people whose work contributed greatly to the 'beauty and presence' of the building: the late Mr A. M. Wilson, retired architect, who designed the Memorial and supervised the work of its construction and decoration; and the late Mr Charles Florence, former Principal Teacher of Art at Buckie High School, who designed the five stained glass windows that depict the sea, fishing and the surrounding community, and light the interior of the Memorial like brightly coloured jewels.

The Buckie Seamen's Memorial. *Michael Charman*

BUCKIE DISTRICT FISHING HERITAGE

Discover the unique fishing heritage of this lovely part of the Moray Firth in informal settings, with the extensive range of photographs, model boats and hundreds of other fishing artefacts in our Heritage Cottage. Watch a fishing video, or listen to one of our many taped conversations with individuals connected with the fishing industry.

The Heritage Cottage is run by local people who give up their free time to preserve the character, history and traditions of the fishing and boatbuilding industries in the Buckie district.

Over the years we have built up an unrivalled database of the history of locally built boats from the latter part of the 19th century to the present day. The information can be quickly accessed by giving a particular boat's name or number, and is supplied free of charge to any interested party.

For further information, please visit www.buckieheritage.org

The *Bezaleel* – its Biblical name means 'under the shadow and protection of God – decorated for a local gala week. *From the collection of the Buckie District Fishing Heritage Museum*

Index